Gauge Invariance Approach to Acoustic Fields

Woon Siong Gan

Gauge Invariance Approach to Acoustic Fields

Springer

Woon Siong Gan
Acoustical Technologies Singapore Pte Ltd.
Singapore, Singapore

ISBN 978-981-13-8753-1 ISBN 978-981-13-8751-7 (eBook)
https://doi.org/10.1007/978-981-13-8751-7

This Springer imprint is published by the registered company Springer Nature Singapore Pte Ltd.
The registered company address is: 152 Beach Road, #21-01/04 Gateway East, Singapore 189721, Singapore

To Raymond Stephens
My Ph.D Supervisor

Foreword

I first encountered Woon Siong Gan during the 22nd International Conference of Sound and Vibration (ICSV22) in Florence in July 2015. After the first meeting, I soon realized that his expressed interests in nonlinear acoustics, fundamental physics and Lie groups were the surface tell tales of wide-ranging curiosity and a talent for careful analysis. He was, and is, a researcher interested in building, but he was, and is, equally interested in ideas. We both were elected as directors of the International Institute of Acoustics and Vibrations (IIAV) for the period 2018–2022, the first time for me and the second time for Woon Siong, who was first elected in 2002.

Not long thereafter, Woon Siong asked if I would be willing to co-chair a special session in "Nonlinear Acoustics and Vibrations" during ICSV23. My first reaction was to accept with pleasure this proposal since I consider nonlinear science as a new workhouse for improving technologies and knowledge in various domains of applied acoustics such as nondestructive testing (NDT), medical imaging and analysis of complex data in the wide sense. Our plan was to invite eminent scientists who are making major contributions in different areas of nonlinear acoustics, to deliver formal talks on their visions of the great challenges and possible pay-offs from Nonlinear Sciences, Technologies, Engineering and Mathematics (Nonlinear STEM) for the twenty-first century. Since 2016, we co-chair the "Nonlinear Acoustics and Vibrations" session in annual ICSV meetings.

Of course, my personal scientific education conducted in the field of theoretical physics to the Department of Nonlinear Physics of the University of Dijon, France, motivated me to accept the invitation to write the foreword for Woon Siong's book. In this topic, I love the idea to associate symmetries and invariance to a new way to extract new suitable parameters able to provide new acoustical information. Since there are several situations in acoustics which are consequences of the symmetry properties of acoustic fields, such as time reversal symmetries and acoustic metamaterials, I have personally made some contributions with my research group where symmetries enable simplifications of the complexity. Since the approach of this book comes from the fundamental mathematics of gauge theory such as change of variables in coordinate systems, symmetry analysis using Lie groups is used, I consider that Woon Siong Gan is the unique researcher enable to exhibit applications of gauge invariance in acoustic metamaterials, nonlinear time reversal acoustics, acoustic diodes, electron–phonon interaction, spintronics, etc. I deeply consider that Woon Siong Gan is uniquely qualified to write this book which associates acoustics and gauge invariance coming from theoretical physics. Considering the huge and wideband qualifications of Woon Siong Gan related to theoretical physics, he is uniquely qualified to convince acousticians about the real added value of his ideas presented in the book, and the benefits readers will get from reading it.

In the past few years, nonlinear sciences has shown to provide several applications, not only in the acoustic fields, but also in electromagnetic, biomedical and neuromorphic sciences. Nonlinear acoustics has gone from an arcane subject known to a few of my students in engineering sciences, to a subject engaging the curiosity of many different research communities including researchers in material characterization, signal and image processing, NDT integrity engineering, biomedical engineering and computer sciences. A major reason for this interest is that nonlinear acoustics and fundamental physics really work. Nonlinear science and group theory offer a robust framework that can exploit massively parallel architectures and applied to practical applications, and they provide a new route towards an understanding of small signals that can be extracted from environmental acoustic noise. Woon Siong Gan book provides a turnpike into this territory.

I consider this book as a keystone for going ahead for two major reasons:

- First, all the main themes using acoustics have been revisited with the gauge invariance approach.
- Second, important contribution of this book was to correct choice of chapter ordering. Important number of examples demonstrated also the real emergence of applications for nonlinear acoustics, metamaterials, etc.

In recognition of his numerous contributions to the board fields of nonlinear acoustics, I hoped that this book would provide the most appropriate reference for our community of acoustics and vibrations. I believe that the throughout ideas and positions to be advocated by these chapters will help to identify important but tractable research directions in nonlinear acoustics and to provide leadership and inspiration to future generations of researchers for solving many fundamental problems from such multidisciplinary areas as NDT, medical imaging and fundamental of acoustics.

The majority of the target audience of this book would be scientists and engineers. However, I also recognized that much of the visions to be projected by this book would most likely call for generations of young blood in the coming decades. Consequently, I would suggest to students in theoretical acoustics to investigate deeply all the ideas presented in this book. I hope all of the students will take advantage of this privileged opportunity and that most of them will apply, in future, the concepts discussed in this book and contribute to enhance them, remembering, at the same time, that fundamental mathematics is necessary for understanding our complex world. This book, together with previous ones written by Woon Siong Gan, will provide a solid foundation of modern acoustics using gauge invariance and will stimulate intense worldwide research activities.

One cannot be around Woon Siong Gan for long without being infected by his enthusiasm, knowledge and energy. That enthusiasm comes across this book. It is also an embodiment of his passion for clear explanations and carefully worked examples. His book does an exceptional job of making the methods easily applicable to acoustic fields available to a wide audience. Woon Siong is deeply interested in the intellectual problem of the conceptualization of acoustic science, but he is interested even more in seeing these systems used in the domain of NDT, metamaterials, etc. This book, I think, will be instrumental in realizing that ambition. I wish to take this opportunity to thank Woon Siong for giving me the opportunity to write these words related to this inspiring book, containing visionary ideas.

I believe that the thoughtful ideas and positions to be advocated by these chapters will help to identify important but tractable research directions in nonlinear acoustics and to provide leadership and inspiration to future generation of acousticians for solving many fundamental problems from such multidisciplinary areas such as metamaterials, eco-acoustics, noise in complex systems, using the powerful tools of neural networks, artificial intelligence, nonlinear dynamics, Lie group theory and advanced signal and image processing. As a tribute to Woon Siong Gan's legacy to acoustics, I remark that he has always been a strong advocate of mathematical rigour as an essential standard and element in the discipline of nonlinear acoustics, even though he has also emphasized the crucial importance of physical interpretations and experimental confirmations.

Blois, France
July 2019

Dr. Serge Dos Santos
Hab. Dir. Rech., Director of the IIAV
Full Member of Academia NDT International

Contents

About the Author

Dr. Woon Siong Gan completed his Ph.D at the age of 24, and is one of the founders of condensed matter physics. He was the first to introduce transport theory into condensed matter physics, and coined the term "transport theory" in 1966, during his Ph.D studies at the physics department of Imperial College London. Today, transport theory is the key foundation of the theoretical design of new materials and enjoys a status in condensed matter physics similar to that of Yang Mills theory in particle physics. His Ph.D thesis (1969) employed a statistical mechanics approach to describe phase transition and ultrasound propagation in semiconductors in the presence of high magnetic fields and low temperatures instead of the usual method of electron-phonon interaction from many body theory. Accordingly, his thesis also played a part in the founding of condensed matter physics.

In 2007 he introduced a gauge invariance approach to acoustics fields, a milestone in the history of gauge invariance in physics. Recently he has also introduced curvilinear spacetime for the treatment of nonlinear acoustics; a suitable choice due to the curved paths of high-intensity acoustic fields.

Dr. Woon Siong Gan authored the books *Acoustical Imaging: Techniques and Applications for Engineers*, published by John Wiley & Sons in 2012, and *New Acoustics Based on Metamaterials*, published by Springer in 2018. This is his third book, *Gauge Invariance Approach to Acoustics Fields*, published by Springer. His current research interests are in transport theory approaches to phase transition, quantum acoustic metamaterials, and curvilinear spacetime approaches to nonlinear acoustics.

Chapter 1
Introduction

Abstract The gauge invariance approach is introduced to acoustic fields. This manifested as symmetry properties of acoustic fields which enable the deeper understanding of the acoustic fields and several important developments in the theories of acoustics and important novel applications.

There are several situations in acoustics which are in fact consequences of the symmetry properties of acoustics fields. Examples are time reversal symmetry, acoustic metamaterials, Goldstone mode, etc. By explaining them in terms of the symmetry principle will enable one to have a deeper and more fundamental understanding of acoustic fields and to explore and discover new phenomena and new materials which are related to the symmetry properties of the acoustic fields.

Symmetry is an important subject in physics. Richard Feynmann mentioned that all fundamental equations of physics have symmetry properties. The study of symmetry is known as gauge theory, and the approach used is known as the gauge invariance approach. Gauge theory was originally proposed by Weyl [1]. Examples of gauge theory in physics are Einstein's general theory of relativity, Yang Mills theory of the standard model of particle physics and the Higgs theory of particle mass.

The symmetry property of fields will enable simplification in complexity. An example is the elastic properties of an isotropic solid. Originally, the elastic constant tensor has nine components. With symmetry properties of an isotropic solid, they will be reduced to only three components, the Lamé constants.

Symmetry is a big topic in physics. Hence, this book covers this subject in fourteen chapters. Both infinitesimal amplitude sound wave and finite amplitude sound wave are covered. It also covers the partial symmetry phenomenon of spontaneous symmetry breaking (SSB). The applications of symmetry are tremendous, from fluids to solids, from electromagnetic field to acoustic field and to the study of cosmos and the universe.

The approach of this book starts with the fundamental mathematics of gauge theory such as the coordinate systems, the Lie group, the vector potential and the covariant derivative. Then all the fundamental properties of gauge invariance such as global gauge invariance, local gauge invariance, spontaneous symmetry breaking.

W. S. Gan, *Gauge Invariance Approach to Acoustic Fields*,
https://doi.org/10.1007/978-981-13-8751-7_1

This book is divided into six sections so as to cover the whole spectrum of physics. There will be amplitude sound waves, gauge invariance for finite amplitude sound waves, gauge invariance in flat spacetime (Minkowski spacetime) and gauge invariance in curvilinear spacetime.

Reference

1. Weyl, H.: Elektron und gravitation. I. Z. Phys. **56**, 330 (1929)

Chapter 2
History of Gauge Theory

Abstract The roots of gauge invariance are given followed by the role of the key gauge field, the vector potential. The milestones of the development of gauge invariance principle in physics are described. The roles of Maxwell, V. Fock, Herman Weyl are provided. Then the Yang–Mills theory, the Glashow–Weinberg–Salam theory and Gell-mann's theory are briefly introduced. Woon Siong Gan's introduction of gauge invariance to acoustic fields equation is considered as a milestone in the theory of gauge invariance in physics. The gauge invariance in acoustic field equation in curvilinear spacetime and the application of gauge invariance to gravitational wave are briefly introduced.

2.1 The Roots of Gauge Invariance

Gauge invariance is a very important principle in physics. The underlying theory of the standard model of particle physics, quantum electrodynamics, quantum chromodynamics, theory of superconductivity, theory of superfluids and the general theory of relativity are all gauge theories. The oldest gauge theory in physics is Maxwell's equations of electromagnetic wave equation. The gauge conditions in Maxwell's equations are:

$$\mathbf{A}' = \vec{A} + \nabla\chi \tag{2.1}$$

$$\phi' = \phi - \frac{1}{c}\frac{\partial\chi}{\partial t} \tag{2.2}$$

where $\vec{A}$ = vector potential, ϕ = scalar potential and χ = gauge function. Both $\vec{A}$ and ϕ are gauge fields.

Equations (1.1) and (1.2) are conditions for the gauge transformations in Maxwell's equations of electromagnetic theory. The concept of gauge invariance arose due to the discovery that different forms of the vector potential $\vec{A}$ produce the same observable forces. The vector potential $\vec{A}$ is the key gauge field. This means that Maxwell's electromagnetic equations are invariant under the gauge transformations

W. S. Gan, *Gauge Invariance Approach to Acoustic Fields*,
https://doi.org/10.1007/978-981-13-8751-7_2

of Eqs. (2.1) and (2.2). This is an Abelian group with the group symmetry U(1). These are gauge invariance applied to classical physics.

After the arrival of quantum mechanics, an additional gauge condition was added on by Fock [1] in 1926:

$$\Psi' = \psi \exp(ie\chi/\hbar c) \tag{2.3}$$

where ψ = wave function, e = electronic charge, $\hbar$ = Planck's constant and c = light velocity.

This means that in order to maintain the same form of the Schrodinger equation, the wave function has to be multiplied by the spacetime-dependent local phase.

Subsequently in 1929, Weyl [2] combined gauge transformations of (2.1), (2.2) and (2.3) to produce the concept of the invariance of a theory known as gauge invariance or gauge symmetry. The name gauge invariance was introduced by Weyl [2]. The gauge symmetry of the quantum version of the electromagnetic theory, known as quantum electrodynamics, is Abelian symmetry described by the U(1) group. The first work on non-Abelian gauge symmetry SU(2) × U(1) was due to Klein [3] in 1938 applying to the electromagnetic and weak interactions. In 1954, Yang and Mills [4] rediscovered non-Abelian gauge symmetry and applied to strong interactions. In the 1960s, there was the Glashow, Weinberg and Salam [5–7]'s paper on the non-Abelian electroweak theory. The discovery in CERN in 1983 of the heavy W and Z bosons confirmed the correctness of the electroweak theory. In the 1970s, a non-Abelian gauge theory for strong interaction of quarks and gluons was developed [5]. One of its creators Gell-Mann [8] gave the theory the name of quantum chromodynamics (QCD) and was based on the SU(3) group. QCD and the electroweak theory form the basis of the standard model which accounts for all the three main forces of nature: strong, electromagnetic and weak forces with the exception of the fourth force the gravitational force.

Hence, gauge invariance plays a key role in physics. Gauge invariance is the underlying principle of the current standard model of strong and electroweak interactions.

2.2 The Vector Potential $\vec{A}$

The vector potential $\vec{A}$ is the key gauge field. Vector potential was introduced by Ludvig V. Lorenz in 1867. By the beginning of the twentieth century, due to the contributions by Heaviside, Lorentz, Clausius and Hertz, on localized charges in motion forming currents, the role of the scalar and vector potentials and their interaction with charged particles, the formal structure of electromagnetic theory and the concept of gauge transformations were founded. In the 1920s with the arrival of quantum mechanics, the interaction of charged particles with a time-varying electromagnetic fields, the issue of the arbitrariness of the electromagnetic potentials arose. The study of the consequences of the change in the electromagnetic potentials on the quantum mechanical wave function is transformed into a general principle that defines quantum gauge fields.

Faraday discovered electromagnetic induction in 1839. This means relative motion of a magnet near a closed circuit induces a momentary flow of current. This started the close link between electric and magnetic fields. Franz E. Neumann in 1845 and 1847 analysed electromagnetic induction in one circuit due to the relative motion of nearby circuits and magnets. He is credited to invent the vector potential. His formula for the induced current allowed one to sense the vector potential:

$$\text{Magnetic interaction energy } W = \frac{I}{c} \oint_C n^o \boldsymbol{A}' \, \mathrm{d}S \tag{2.4}$$

with

$$\boldsymbol{A}'(x) = \frac{I'}{c} \oint_{C'} \frac{n'}{r} \mathrm{d}S' \tag{2.5}$$

where C, C' = circuits, A' = vector potential of the current I' flowing in circuit C', r = distance = $x - x'$ where x and x' are coordinates of $\mathrm{d}S = n\mathrm{d}S$ and $\mathrm{d}S' = n\mathrm{d}S$, I and I' = currents and $\mathrm{d}S$, $\mathrm{d}S'$ = elements.

In 1857, Kirchhoff showed the relation between the scalar potential ϕ and the vector potential $\vec{A}$:

$$\vec{A} = \frac{1}{c} \frac{\partial \phi}{\partial t} \tag{2.6}$$

2.3 Milestones of the Development of Gauge Invariance Principle in Physics

2.3.1 *Maxwell's Role in Developing the Gauge Theory*

Maxwell developed the electromotive force as:

$$c\boldsymbol{E} = -\frac{\mathrm{d}\vec{A}}{\mathrm{d}t} = -\frac{\partial \vec{A}}{\partial t} + (\boldsymbol{v} \cdot \nabla)\vec{A} \tag{2.7}$$

and the gauge transformation:

$$\boldsymbol{A}' = \vec{A} - \nabla \chi \tag{2.8}$$

In the meantime, Lorentz mentioned that his retarded potentials $\vec{A}$ and ϕ are solutions of Maxwell's equations and that the following gauge condition must be satisfied:

$$\nabla \cdot \vec{A} + \frac{1}{c^2}\frac{\partial \phi}{\partial t} = 0 \tag{2.9}$$

This is now known as the Lorentz gauge condition.

2.3.2 *The Role of Fock [1] in the History of Gauge Invariance [1926b]—The Quantum Era*

1926 saw the arrival of quantum mechanics, and the emergence of the principle of gauge invariance in quantum region. Fock [1] discovered the symmetry under gauge transformation in equations (2.1), (2.2) and (2.3) of the quantum mechanical system of a charged particle interacting with electromagnetic fields.). His paper was submitted on 30 July 1926 and published on 2 October 1926. This means that the Schrödinger equation is invariant under the gauge transformations (2.1), (2.2) and (2.3). He showed that the equations are invariant under the change in the vector and scalar potentials provided that the gauge transformation condition in (2.3) is fulfilled. Also in this paper, he first discussed the special-relativistic wave equation of his earlier paper with electromagnetic interactions and addressed the effect of the change in the vector and scalar potentials (2.1 and 2.2). He went on to treat a five-dimensional general relativistic formalism, similar to but independent of Klein [2]. Also he added the following note after the proof of this paper: "While this note was in proof, the beautiful work of Oskar Klein [published on 10 July] arrived in Leningrad," and that the principal results were identical.

The idea of an Abelian gradient transformation was first conceived by Fock [1] in 1926. He also discovered that the electromagnetic interaction of charged particles has a gradient invariance in the framework of quantum mechanics. These transformations and invariance were respectively named Eichtransformation and Eichinvarianz by Weyl [2] in 1929 (the German verb *zu eichen* means *to gauge*).

2.3.3 *The Role of Hermann Weyl [2] in the History of Gauge Invariance [1929]*

In 1929, shortly after the formulation of QED, Weyl [2] wrote a paper relating electromagnetism to matter field. He discovered the equations remain invariant when one makes simultaneous substitutions $\psi \rightarrow \Lambda\psi$.fp $\rightarrow$ fp $- \partial\lambda/\partial xp$. In the abstract of this important paper [2]. He states "The Drac field equation for ψ together with the Maxwell's equations for the four potential fp of the electromagnetic field have an invariant property. It seems that this new principle of gauge invariance, which follows not from speculation but from experiment, tells us that the electromagnetic field is a necessary accompanying phenomenon not of gravitation, but of material wave field represented by ψ." From this paper, the concept of gauge invariance was born. Weyl's route to gauge invariance illustrates the fact that theoretical proposals must be verified by experiments.

Historically, Weyl [2]'s 1929 papers were a watershed. They enshrined fundamentally the modern principle of gauge invariance, in which the existence of the four-vector potentials (and field strengths) follow from the requirement of the invariance of the matter equations under gauge transformations such as (2.1), (2.2) and (2.3) of the matter fields. Gauge invariance is the touchstone of the theory of gauge fields, so dominant in theoretical physics in the second half of the twentieth century.

2.3.4 The Yang–Mills Theory

The Yang–Mills theory is a gauge theory based on the SU(N) group. In 1954, Chen Ning and Robert Mills [4] extended the concept of Abelian group, U(1) group for quantum electrodynamics to non-Abelian groups to explain strong interactions. They developed their theory without realizing that the Ph.D. student Shaw [9] under Abdus Salam at Cambridge University also developed the same at the same time. Due to the requirement of particles must be massless to fulfil gauge invariance, as criticized by W. Pauli, the theory did not receive attention till 1960 with the arrival of spontaneous symmetry breaking when the concept of particles acquiring mass through symmetry breaking was put forward, initially by Goldstone [10], Nambu and Jona-Lasinio [11–13].

This rekindled the interest in the Yang–Mills [4] theory studies which proved successful in the formulation of both electroweak unification and quantum chromodynamics (QCD). The electroweak interaction is described by SU(2) $\times$ U(1) symmetry breaking non-Abelian group, while QCD is an SU(3) non-Abelian group Yang–Mills theory. The electroweak theory is the combination of SU(2) group with U(1) group. Here quantum electrodynamics (QED) is described by the U(1) group, and replaced in the unified electroweak theory by a U(1) group representing a weak hypercharge rather than electric charge. The massless bosons from the SU(2) $\times$ U(1) electroweak theory interact after spontaneous symmetry breaking to produce the photon field and the three massive weak bosons. The standard model combines the unified electroweak interaction, unifying the weak and electromagnetic interaction with the strong interaction through the non-Abelian symmetry group SU(2) $\times$ U(1) $\times$ SU(3). Experimentally, the unification of strong, electromagnetic and weak interactions is shown by the observation of the running of the coupling constants believed that they all converge to a single value at very high energies.

The first manifestation of the physical application of the Yang–Mills theory is the Weinberg–Salam model [6, 7], a SU(2) $\times$ U(1) Yang–Mills theory, of electroweak theory unifying electromagnetic interaction with weak interaction where the gauge fields are identified as the photon fields with the hypothetical massive vector mesons mediating weak interactions. The Yang–Mills [4] theory was developed to describe strong interactions. The significance of this theory is that it is the only model which can account for asymptotic freedom [14, 15] which means interaction forces become negligible at short distances.

Subsequently, the SU(3) Yang–Mills [4] theory [5] also known as quantum chromodynamics, coupled to quarks, provides the realistic theoretical framework for the MIT/SLAC experiments on high-energy lepton–nucleon scattering. These show the physical significance of the non-Abelian gauge fields which although are not identified with the observed particles, they provide the glue that keeps the quarks bound strongly inside hadrons so strongly to be permanently confined.

2.3.5 *The Glashow-Feinberg-Salam's Electroweak Theory*

In 1960s, Glashow, Weinberg and Salam [5–7] developed the non-Abelian SU(2) × U(1) symmetry group theory to unify the electromagnetic interaction and the weak interaction. The unification is accomplished under a SU(2) × U(1) gauge group. The verification of the theory was experimentally established in two stages: the first being the discovery of neutral currents in neutrino scattering by the Gargamelle collaboration in 1973, and the second in 1983 by the UA1 and the UA2 collaborations involving the discovery of the W and Z gauge bosons in proton–antiproton collisions at the converted Super Proton Synchrotron. In 1999, Gerardus ‘t Hooft and Veltman [16, 17] were awarded the Nobel Prize for demonstrating that the electroweak theory is renormalizable.

2.3.6 *Gell-Mann's Quantum Chromodynamics for Strong Interaction*

Quantum chromodynamics (QCD) is a gauge theory based on the non-Abelian SU(3) group. It is the theory of strong interaction between quarks and gluons. Murray Gell-Mann [8] coined the word quark. Quarks and gluons are the fundamental particles that make up the composite hadrons like pion, neutron and proton. In QCD, the word colour is a property in QCD which is the analog of electric charge in QED. Gluons are the equivalent of protons in QED. They are the force carriers of the strong interaction force theory, like photons the carriers of electromagnetic force in QED. QCD is an important part of the standard model. Over the years, there are a large number of experiments verifying the theory.

Every field theory of particle physics has symmetry properties whose existence is deduced from observations. Symmetries can be divided into:

- local symmetries that act independently at each point in spacetime. Each such symmetry is the basis of a gauge theory that requires the introduction of its own gauge bosons.
- global symmetries, which are symmetries whose operations must be applied simultaneously to all points of spacetime. An example of global symmetry is reflection symmetry

QCD is a gauge theory of the SU(3) non-Abelian gauge group obtained with the colour charge defining a local symmetry. Since the strong interaction does not discriminate between different flavours of quark, QCD has approximate flavour symmetry, broken by the differing masses of the quarks.

2.3.7 W. S. Gan's Introduction of Gauge Invariance to Acoustic Wave Equation of Motion

In 2007, Gan [18] introduced gauge invariance or gauge theory into acoustic wave equation of motion. This will enable a deeper understanding of the symmetry properties of acoustic fields and new phenomena to be discovered in acoustics. The starting point is due to the several similarities between electromagnetic waves and acoustic waves due to the common wave phenomena. The introduction of gauge invariance into electromagnetic waves has enabled discoveries of several important theories such as the Yang–Mills theory, electroweak theory and quantum chromodynamics. Likewise for acoustic waves, the introduction of gauge invariance should give rise to important new theories in the future and bring acoustics to the highest arena. Gan [18] introduced the two gauge conditions (1.1) and (1.2) from electromagnetic wave equation into acoustic wave equation:.

Gan [18] extended the gauge invariance property of Maxwell's equation for electromagnetic theory to the acoustic field equations. Gauge invariance which includes symmetries is a basic property of field theory which covers strong nuclear forces, electromagnetic force and gravitational force. In extending gauge invariance approach to acoustic fields, it will be a more sophisticated approach than the vector theory of acoustic fields. We address the symmetry properties of the acoustic field equations, the application of gauge invariance to negative refraction and interpretation of the inhomogeneous wave equation in terms of gauge invariance. Gauge invariance has long been applied to electromagnetic wave theory in Maxwell's equations. Due to the similarities between electromagnetic waves and acoustic waves, as both are wave phenomena, the interpretation of acoustic fields in terms of gauge invariance will provide more understanding of the acoustic fields and throw lights on new potential applications.

The applications of gauge invariance to Maxwell's equation has given rise to several important developments on standard model of the particle physics and to superconductivity and superfluidity. The introduction of gauge invariance to the acoustic wave equation of motion on the other hand will find applications in condensed matter physics. An example is the acoustic metamaterial which is an outcome of the symmetry properties of the acoustic field. Also gauge invariance can be introduced in electron–phonon interaction with application in spintronics [19]. Another example is the application of gauge invariance to electron–phonon interaction that has produced the theory of superconductivity which is a gauge theory. Also gauge theory can be applied to acoustic multiple scattering problem in diffraction tomography. Other applications are in time reversal acoustics and the interpretation of phonon as a Goldstone mode.

In electromagnetic theory, the Lorentz gauge condition is given as

$$\nabla \cdot \vec{A} + \frac{1}{c^2}\frac{\partial \phi}{\partial t} = 0 \tag{2.10}$$

With Lorentz condition, the acoustic wave equation can be expressed in terms of the scalar and vector potentials as:

$$\nabla^2 \vec{A} - \frac{1}{V_s^2}\frac{\partial^2 \vec{A}}{\partial t^2} = 0 \tag{2.11}$$

$$\nabla^2 \phi - \frac{1}{V_l^2}\frac{\partial^2 \phi}{\partial t^2} = 0 \tag{2.12}$$

where $V_l = (c_{11}|\rho)^{1/2}$.

This shows that the acoustic fields T and V can be expressed in terms of the gauge fields $\vec{a}$ and ϕ. This is a more accurate method of obtaining solutions for the acoustic fields.

The Lorentz gauge condition has the advantage of introducing complete symmetry between the scalar and vector potentials, i.e. it makes both potentials satisfy the same wave equation as that obeyed by the fields. Equations (2.11) and (2.12) are a symmetric set of equations.

By using the analogy of momentum density P as equivalent to B and stress field T as equivalent to $\vec{E}$, one has

$$\vec{T} = -\nabla \phi - \frac{\partial \vec{A}}{\partial t} \tag{2.13}$$

and

$$P = \nabla \times \vec{A} \tag{2.14}$$

By inspection of Eqs. (2.13) and (2.14), one finds that the resultant stress field and momentum density field are unchanged by transformation of the following types which are gauge conditions:

$$\mathbf{A}' = \vec{A} + \nabla \chi \tag{2.15}$$

$$\phi' = \phi - \frac{1}{c}\frac{\partial \chi}{\partial t} \tag{2.16}$$

shown in (2.10) and in (2.2) where ϕ is a function of the coordinates and the time. This means that if any physical law involving elastic energy interactions is to be expressed in terms of the general elastic energy potential $A\sim$ and ϕ, then such a physical law must be unaffected by a transformation of that given by Eqs. (2.15) and (2.16). These transformations are usually known as gauge transformations, and a physical law that is invariant under such a transformation is said to be gauge invariant. The property of gauge invariance ensures that the physical law will not lead to consequences that cannot be expressed in the field formulation of the interaction of elastic properties.

From the above field theory approach, we find that both the vector fields $\vec{V}$ and $\vec{T}$ can be expressed in terms of a scalar potential and a vector potential.

The application of gauge invariance approach to acoustic fields to negative diffraction:

Veselago [20] derived the negative refraction theory from the consideration of negative permittivity and negative permeability, and is limited only to isotropic materials and electromagnetic wave. Here one extends his theory to acoustic wave and to anisotropic materials such as piezoelectric materials. Gauge invariance is used in this study. It is well known that an important property of gauge invariance is symmetry.

In an anisotropic material, the compliance and stiffness possess rotational symmetry. Compliance and stiffness together describe the intrinsic elastic properties of the medium. The compliance constant describes the elastic properties of a medium in a manner analogous to the description of its electrical properties by the permittivity matrix elements. If the medium itself is symmetric with respect to a particular transformation of coordinates, then the compliance and stiffness matrices must themselves be unchanged by the same transformation. Symmetries for anisotropic media are much more complicated than for the isotropic case. When rotating in the clockwise direction, it gives rise to left-handed phenomenon such as the negative refraction. When rotating in the anticlockwise direction, it gives rise to the right-handed phenomenon such as positive refraction. Due to rotational symmetry, both the right-handed phenomenon and left-handed phenomenon satisfy the acoustic field equations. The stress field, the velocity field and the acoustic Poynting vector together form a right-handed triplet or a left-handed triplet depending on the direction of energy flow or the direction of the Poynting vector. According to parity conservation, acoustic law at the deepest level, there is no differentiation of right-handed and left-handed treatments. The performance of an object and that of its mirror image will satisfy the same law of physics. The negative refraction in fact is a mirror image of the positive refraction.

2.3.8 Gauge Invariance in Acoustic Wave Equation in Curvilinear Spacetime

The above gauge invariance treatments from Sects. 2.3.1 to 2.3.7 are all meant for the flat spacetime or the Minkowski spacetime. Here one extends the treatment of gauge invariance in acoustic wave equation to curvilinear spacetime. The first step is one has to start with the relativistic acoustic wave equation for curvilinear spacetime and express it in terms of the vector potential and the scalar potential.

2.3.9 Application of Gauge Invariance to Gravitational Wave

Gravitational wave was experimentally detected in 2016 [21] and awarded the 2017 Nobel Prize in Physics. This is the third form of wave in nature after electromagnetic wave and acoustic wave with entirely different nature and properties from the other two types of waves. Hence, the introduction of gauge invariance into this form of wave will discover several interesting novel theories and phenomena. The obvious theoretical platform for the treatment of gravitational wave will be the curvilinear spacetime.

References

1. Fock, V.: Uber die invariante Form der Wellen- und der Bewegungsgleichungen f ù ur í einen geladenen Massenpunkt. Zeit. fur Physik 3 í **9**, 226–232 (1926) [subm. 30 July 1926, publ. 2 October 1926]
2. Weyl, H.: Elektron und gravitation. I. Z. Phys. **56**, 330 (1929)
3. Klein, O.: On the theory of charged fields. In: New Theories in Physics, Conference Organized in Collaboration with the International Union of Physics and the Polish Intellectual Cooperation Committee, Warsaw, May 30th–June 3rd, 1938
4. Yang, C.N., Mills, R.L.: Conservation of isotopic spin and isospin gauge invariance. Phys. Rev. **96**, 191–195 (1954)
5. Glashow, S.: The renormalizability of vector meson interactions. Nucl. Phys. **10**, 107 (1959)
6. Salam, A., Ward, J.C.: Weak and electromagnetic interactions. Nuovo Cimento. **11**(4), 568–577 (1959)
7. Weinberg, S:. A model of leptons. Phys. Rev. Lett. **19**, 1264–1266 (1967)
8. Gell-Mann, M.: The eightfold way: a theory of strong interaction symmetry (No. TID-12608; CTSL-20). California Institute of Technology, Pasadena. Synchrotron Lab (online) (1961)
9. Shaw, R: Invariance under general isotopic gauge transformations. Cambridge University Ph.D. thesis, unpublished (1955)
10. Goldstone, J.: Field theories with "superconductor" solutions. Il Nuovo Cimento. **19**, 154–164 (1961)
11. Nambu, Y.: Quasiparticles and gauge invariance in the theory of superconductivity. Phys. Rev. **117**, 648–663 (1960)
12. Nambu, Y., Jona-Lasinio, G.: Dynamical model of elementary particles based on an analogy with superconductivity. I. Phys. Rev. **122**, 345–358 (1961)

13. Nambu, Y., Jona-Lasinio, G.: Dynamical model of elementary particles based on an analogy with superconductivity. II. Phys. Rev. **124**, 246–254 (1961)
14. Gross, D., Wilczek, F.: Phys. Rev. Lette. **30**, 1343 (1973)
15. Politzer, D.: Phys. Rev. Lett. **30**, 1346 (1973)
16. 't Hooft, G.: The renomalization procedure for Yang-Mills fields. Ph.D. Thesis, Utrecht University (1972)
17. 't Hooft, G., Veltman, M.: Regularization and renormalization of gauge fields. Nucl. Phys. **B44**, 189 (1972)
18. Gan, W.S.: Gauge invariance approach to acoustic fields. In: AkiYAMA, I. (ed.) Acoustical Imaging, vol. 29, pp. 389–394. Springer, The Netherlands (2007)
19. Dartora, C.A., Cabrera, G.G.: The electron–phonon interaction from fundamental local gauge symmetries in solids. J. Phys. A: Math. Theor. **47**(3) (2013)
20. Veselago, V.G.: The electrodynamics of substances with simultaneously negative values of ε and μ. Sov. Phys. Usp. **10**(4), 509–514 (1968)
21. Abbott, B.P., et al.: Observation of gravitational waves from a binary neutron star inspiral, Phys. Rev. Lett. **119**, 161101 (2017)

Chapter 3
Coordinate Systems as the Framework of Equations

Abstract General covariance is explained. Then, various forms of coordinate systems are introduced: Cartesian coordinate system, cylindrical coordinate system and curvilinear coordinate system. These give rise to the concept of coordinate systems as the framework of equations.

3.1 General Covariance

Richard Feynmann mentioned that all fundamental laws of physics have symmetry properties [1]. This property is represented by equations. The symmetry property of equations is manifested in the form of coordinates. This means that the form of the equations must remain the same subjected to all types of coordinate systems. This is known as general covariance. This shows that coordinate systems will control the requirements of the equations or the coordinate system is the framework or backbone of the equations. The most well-known example of general covariance is given by Albert Einstein's general theory of relativity [2], in which the starting point is to address that all equations of physics must remain unchanged in form when represented by any coordinate system or no change in form subjected to coordinate transformations. Tensor analysis plays an important role in coordinate transformations.

Gauge theory is a field theory in which the equations of motion do not change under coordinate transformations. Gauge transformations are also known as coordinate transformations. In the Preface of Dirac's Quantum Mechanics, he mentioned on transformation theory and method of coordinates and the quantity of fundamental importance, the invariants of transformations.

3.2 Forms of Coordinate Systems

There are several forms of coordinate systems such as Euclidean coordinate system, Cartesian coordinate system, polar coordinate system, spherical coordinate system,

W. S. Gan, *Gauge Invariance Approach to Acoustic Fields*,
https://doi.org/10.1007/978-981-13-8751-7_3

cylindrical coordinate system and curvilinear coordinate system. The following sections will illustrate a few popular coordinate systems.

3.2.1 Cartesian Coordinate System

The Cartesian coordinate system consists of two perpendicularly intersecting lines. The point of intersection is known as the origin. The position of each point is specified by a pair of numerical coordinates. These are the distances to the origin. The two intersecting lines are also known as the coordinate axis of the system (Fig. 3.1).

The above concept has been extended also to three-dimensional case using three perpendicular lines instead. In this case, there will be three Cartesian coordinates. One can extend the same principle to specify the position of any point in three-dimensional space by three Cartesian coordinates. This can be extended even to the n-dimensional case. In this case, there will be n Cartesian coordinates corresponding to distances from the origin to the n mutually perpendicular planes (Fig. 3.2).

As the name implied, Cartesian coordinates were invented by the French mathematician René Descartes in 1637 [4]. However, Pierre de Fermat [3] also discovered the three dimensions, but his discovery was unpublished. The invention of Cartesian coordinates provided the first systematic link between Euclidean geometry and algebraic equations. This caused a revolution in mathematics. One example is the description of geometrical shapes such as curves by algebraic equations. For instance, circle of radius r can be described using the Cartesian coordinates x and y as $x^2 + y^2 = r^2$.

Cartesian coordinates are used in many branches of mathematics such as differential geometry, linear algebra, group theory, complex analysis. Cartesian coordinates

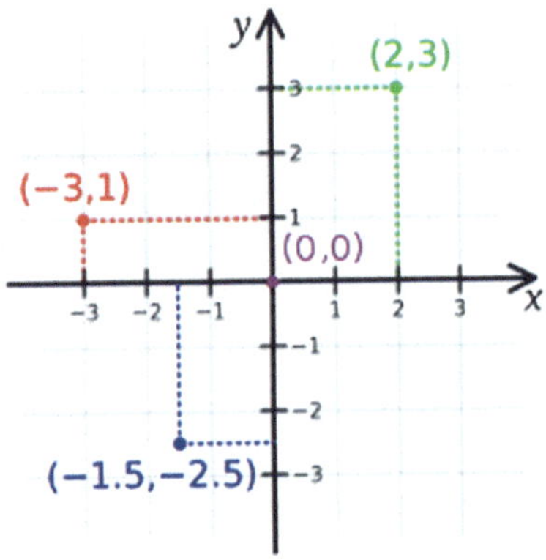

Fig. 3.1 Illustration of a Cartesian coordinate plane. Four points are marked and labelled with their coordinates: (2, 3) in green, (−3, 1) in red, (−1.5, −2.5) in blue and the origin (0, 0) in purple

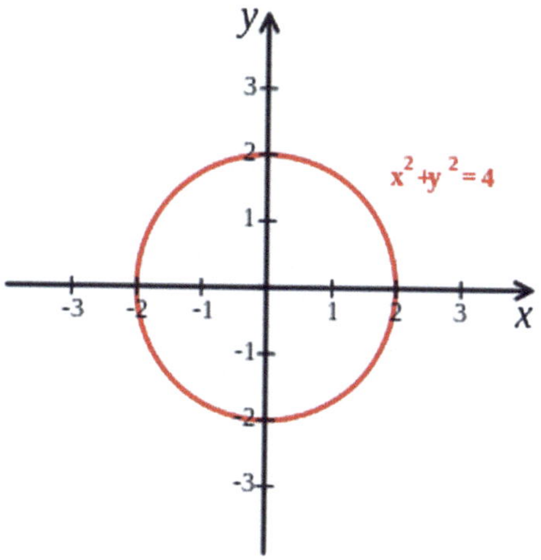

Fig. 3.2 Cartesian coordinate system with a circle of radius 2 centred at the origin marked in red. The equation of a circle is $(x-a)^2+(y-b)^2=r^2$ where a and b are the coordinates of the centre (a, b) and r is the radius

are useful tools in several subjects involving geometry. Examples are physics, engineering and astronomy.

The development in Cartesian coordinate system also led to the subsequent invention of the calculus by Gottfried Wilhelm Leibniz and Isaac Newton [3]. Also the concept of vector spaces [4] is a generalization of the Cartesian coordinate system.

One-Dimensional Cartesian Coordinate System

The simplest example of a one-dimensional Cartesian coordinate system is a straight line. The only issue here is to determine the positive and the negative portions of the line. First, the origin O has to be specified and used as the starting point. Then, the portion to the right will be the positive part and the portion to the left will be the negative part.

Two-Dimensional Cartesian Coordinate System

For the two-dimensional case, first a pair of perpendicular axes have to be defined. Each axis will have a single unit of length. The point of intersection is the origin. For a point P, the two-dimensional coordinates x and y are the horizontal distance and the vertical distance of P from the y-axis and x-axis, respectively. The x-coordinate is also known as the abscissa and the y-coordinate as the ordinate, respectively. A two-dimensional plane containing the x- and y-axes is known as the xy-plane.

Three-Dimensional Cartesian Coordinate System

A three-dimensional Cartesian coordinate system will consist of a triplet of axes. Each axis will have a single unit of length. All the three axes will be perpendicular

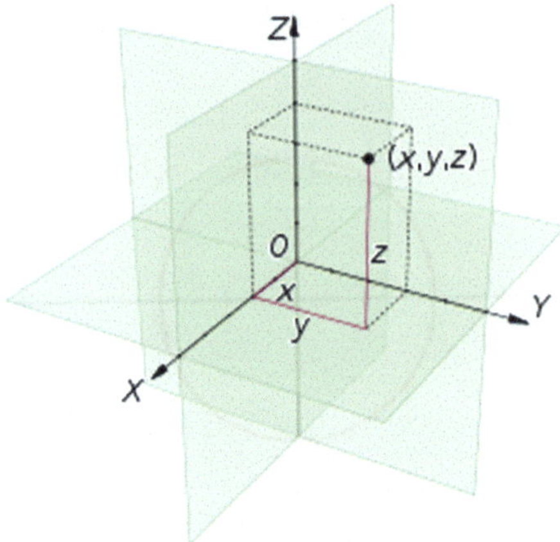

Fig. 3.3 Three-dimensional Cartesian coordinate system, with origin O and axis lines X, Y and Z, oriented as shown by the arrows. The tick marks on the axes are one length unit apart. The black dot shows the point with coordinates $x = 2$, $y = 3$ and $z = 4$ or (2, 3, 4)

to each other. Again as in the two-dimensional case, the three coordinates are given by the perpendicular distances of the point in the three-dimensional space from the x-, y-, z-axes, respectively (Fig. 3.3).

3.2.2 Cylindrical Coordinate System

A cylindrical coordinate system is a three-dimensional coordinate system, but it is different from a three-dimensional Cartesian coordinate system which consists of the x-, y-, z-axes as straight lines, perpendicular to each other. The starting point of the three axes are the origin (Fig. 3.4).

The three coordinates (ρ, φ, z) of a point P are defined as:

- ρ, the axial distance, and is the Euclidean distance from the z-axis to the point P.
- φ, the *azimuth* is the angle between the reference direction on the chosen plane and the line from the origin to the projection of P on the plane.
- z, the axial coordinate or height is the signed distance from the chosen plane to the point P.

Cylindrical coordinates are useful in the description of objects and phenomena with rotational symmetry about the longitudinal axis. Some examples are sound field distribution in a cylindrical object, vibration distribution in a metal cylinder, sound field distribution in an organ pipe and so on.

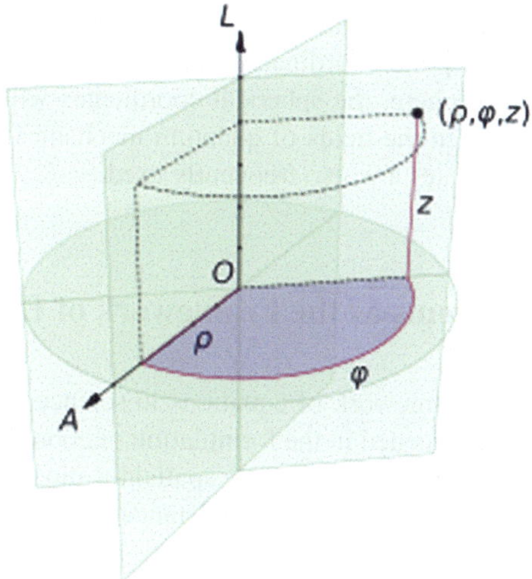

Fig. 3.4 Cylindrical coordinate system with origin O, polar axis A and longitudinal axis L. The dot is the point with radial distance $\rho = 4$, angular coordinate $\varphi = 130°$ and height $z = 4$

3.2.3 *Curvilinear Coordinate System*

Curvilinear coordinate system is a coordinate system in which the coordinate lines may be curved. They include rectangular, spherical and cylindrical coordinates that are curved. These coordinates can be converted from Cartesian coordinates by using a local coordinate transformations, a one-to-one mapping at each point. The name curvilinear coordinates was coined by Lamé, a French mathematician, to describe the fact that the coordinate surfaces of the curvilinear system are curved.

The significance of curvilinear coordinate system is that it fulfils the requirement of general covariance; that is, all physical laws must be independent of any particular coordinate systems used in describing them mathematically if they are to be valid. Also a study of the consequences of this requirement leads to tensor analysis. The formalism of curvilinear coordinates leads to a unified and general description of the standard coordinate systems.

Equations are expressed in terms of physical parameters and quantities, which are represented in terms of coordinate systems. Hence, curvilinear coordinate system as a generalized form of coordinates can be used to describe the behaviours of physical quantities in the forms of scalars, vectors or tensors. These physical parameters used in the mathematics of vector calculus and tensor analysis can be under coordinate transformations to obey general covariance; that is, the equations will become unchanged for any curvilinear coordinate system. The transformation will be subjected to the transformation rules for scalars, vectors and tensors.

The curvilinear coordinate system is simpler to use than the Cartesian coordinate system in a practical application. An example is that for a problem with spherical symmetry such as the motion of particles under central forces will be easier to solve in terms of spherical polar coordinates than in Cartesian coordinates. For the motion of a particle in a sphere, the spherical coordinates will be used instead of the Cartesian coordinates. In the fields of quantum mechanics, relativity and earth sciences, spherical coordinates are also frequently used.

3.3 Coordinate Systems as the Framework of Equations

Coordinate system is the framework of equations in science and engineering. The following requirements are needed in the formulation of coordinate systems: (1) the origin or (0, 0) must be specified for a specific spatial location, (2) the orientation of the axes must be defined and (3) the units of distance which define the spatial size of the coordinates must be specified.

References

1. Feymann, R.: The character of physical law. BBC (1965)
2. Einstein, A.: The foundation of the general relativity. Ann. Phys. **354**(7), 790 (1916)
3. Bix, R.A., D'Souza, H.J.: Analytic geometry. Encyclopaedia Britannica
4. Burton, D.M.: The History of Mathematics/An Introduction, 7th edn, p. 374. McGraw-Hill, New York (2011)

Chapter 4
Gauge Fields

Abstract First, an introduction is followed by the formulation of electric field and magnetic field in terms of gauge fields. Then, the relation between gauge fields and gauge transformation, field strength and gauge fields are provided. Then, the role of gauge field in local gauge invariance is described. Gauge field and phase of the wave function are described. The verification of the physical effect of the gauge field by the Aharanov–Bohm effect is given. Then, gauge fields in acoustics are stated.

4.1 Introduction

The electromagnetic potentials, the magnetic vector potential $\vec{A}$ and the electric scalar potential are known as gauge fields and gauge potentials. They are of more fundamental nature than the electric field $\vec{E}$ and the magnetic field $\vec{B}$. In fact, both $\vec{E}$ and $\vec{B}$ can be expressed in terms of $\vec{A}$ and ϕ. Also, by so doing, the number of six equations for vectors $\vec{E}$ and $\vec{B}$ will be reduced to four equations to the three components of the vector $\vec{A}$ and scalar ϕ. In 2007 [1], Woon Siong Gan [1] introduced gauge invariance from the electromagnetic field equations to the acoustic field equations. The electromagnetic fields $\vec{E}$ and $\vec{B}$ are analogous to the acoustic fields of $\vec{T}$ and $\vec{v}$, the stress field and the particle velocity field. $\vec{A}$ and ϕ were thought to be of the mathematical concept till the proposed experiment of Aharanov and Bohm [2] in 1959 that they are in fact physical entities.

4.2 Formulation of Electric Field and Magnetic Field in Terms of Gauge Fields

The following are two of the four Maxwell's equations:

$$\nabla \vec{B} = 0 \tag{4.1}$$

W. S. Gan, *Gauge Invariance Approach to Acoustic Fields*,
https://doi.org/10.1007/978-981-13-8751-7_4

$$\nabla \times \vec{E} = -\frac{\partial \vec{E}}{\partial t} \tag{4.2}$$

Now, the electromagnetic fields $\vec{B}$ and $\vec{E}$ are usually written in terms of the gauge fields as:

$$\vec{B} = \nabla \times \vec{A} \tag{4.3}$$

and

$$\vec{E} = -\nabla\phi - \frac{\partial \vec{A}}{\partial t} \tag{4.4}$$

4.3 Gauge Fields and Gauge Transformation

The gauge fields have important gauge property. That is, after performing gauge transformation on the gauge fields and substituting them back into Maxwell's equations, there will be no change in the form of Maxwell's equations. That is, it shows the gauge invariance property of Maxwell's equations.

If one chooses these two Maxwell's equations, (4.2) and (4.3) and performs the following gauge transformations,

$$\vec{A}' = \vec{A} - \nabla f(\vec{r}, t) \tag{4.5}$$

and

$$\phi' = \phi + -\frac{1}{c}\frac{\partial f(\vec{r}, t)}{\partial t} \tag{4.6}$$

where f = gauge function, $\vec{A}$ = vector potential, ϕ = scalar potential and c = velocity of light.

With the substitution of (4.5) and (4.6) into (4.3) and (4.4), there is no change in the forms of (4.3) and (4.4).

4.4 Field Strength and Gauge Fields

Field strength is a mathematical concept and it is written in terms of the vector potential, a gauge field. It is a four-dimensional tensor and is given as the derivative of the vector potential as:

$$F_{\mu\nu} = \partial_\mu A_\nu - \partial_\nu A_\mu \tag{4.7}$$

This is a more fundamental approach to field strength instead of expressing it as the square of the magnitude of the electric field or the magnetic field.

4.5 Role of Gauge Field in Local Gauge Invariance

In local gauge invariance, the vector potential will appear in the Lagrangian satisfying local gauge invariance. For globally gauge invariance, the Lagrangian is given by:

$$\text{Lagrangian}_{\text{global}} = \frac{1}{2}(\partial_\mu \phi)^T \partial^\mu \phi - \frac{1}{2} m^2 \phi^T \phi \tag{4.8}$$

where ∂_μ = Einstein notation for the partial derivative of ϕ in each of the four dimensions, ϕ = a vector of fields = $(\phi_1, \phi_2, \ldots \phi_n)^T$ and T^α matrices are generators of the SO(n) group.

To extend to local gauge invariance, one has to replace the partial derivative by the covariant derivative D_μ. The covariant derivative is introduced to preserve the local gauge invariance of the Lagrangian. That is, for the Lagrangian to include the interaction Lagrangian.

$$D_\mu = \partial_\mu - ig\, A_\mu \tag{4.9}$$

where g = coupling constant related to the strength of interaction and A = vector potential

Replacing the ∂_μ by D_μ in (4.8), one has:

$$L_{\text{loc}} = \frac{1}{2}(D_\mu \phi)^T D_\mu \phi - \frac{1}{2} m^2 \phi^T \phi \tag{4.10}$$

Substituting (4.9) into the first term of (4.10), one obtains

$$L_{\text{int}} = i\frac{g}{2}\phi^T A_\mu^T \partial_\mu \phi + i\frac{g}{2}(\partial_\mu \phi)^T A^\mu \phi - \frac{g^2}{2}(A_\mu \phi)^T A^\mu \phi \tag{4.11}$$

The difference between the global gauge invariance and the local gauge invariance is due to this interaction Lagrangian.

4.6 Gauge Field and Phase of the Wave Function

For global symmetry, one has

$$\psi'(\vec{r}, \mathrm{t}) = \psi(\vec{r}, \mathrm{t}) e^{i\lambda(\vec{r}, t)} \tag{4.12}$$

where ψ = wave function and $\boldsymbol{\lambda}$ = phase function.

To extend this to the local phase symmetry, one starts with the Schrodinger's equation:

$$\frac{1}{2m}\left(\vec{p}+\frac{e}{c}\vec{A}\right)^2\psi = (E+e\phi)\psi \tag{4.13}$$

where p = momentum, e = electronic charge, E = electric field and m = mass of electron.

For local symmetry, one has

$$f(\vec{r},\mathrm{t}) = \frac{hc}{2\pi e}\boldsymbol{\lambda}(\vec{r},t), \quad \text{then} \quad \psi'(\vec{r},t) = \psi(\vec{r},t)\mathrm{e}^{-i2\pi e f(\vec{r},t)/hc} \tag{4.14}$$

where h = Planck's constant.

and f is related to $\vec{A}$ as given by (4.5).

This shows that the phase function is related to the gauge field $\vec{A}$.

If a gauge transformation is performed, such that $\vec{A}' = 0$, then $\vec{A}' = \vec{A} - \nabla f(\vec{r}) = 0$ and

$$\vec{A} = \nabla f(\vec{r}) \tag{4.15}$$

$$\int_{\vec{r_0}}^{\vec{r}} \mathrm{d}\vec{r}\,\vec{A} = \int_{\vec{r_0}}^{\vec{r}} \mathrm{d}\vec{r}\,.\nabla\vec{f} = f(\vec{r}) - f\left(\vec{r_0}\right) \tag{4.16}$$

If

$$f\left(\vec{r_0}\right) = 0, \quad \text{then} \quad f(\vec{r}) = \int_{\vec{r_0}}^{\vec{r}} \mathrm{d}\vec{r}\,\vec{A} \tag{4.17}$$

4.7 The Verification of the Physical Effect of the Gauge Field $\vec{A}$ by the Aharanov–Bohm Effect

The Aharanov–Bohm effect [2] is the proposal of an experiment to prove that the gauge fields' magnetic vector potential and the electric scalar potential exist physically and not just mathematical concepts. This is a quantum mechanical phenomenon. It proposed the experimental verification of the physical existence of gauge fields. The Aharanov–Bohm effect [2] was described in their paper in 1959. One year later, in 1960, Robert G Chambers [3] performed the experiment proposed by Aharanov and Bohm and proved the existence of the effect.

The Aharanov–Bohm effect [2] proposed that a particle even when travelling through a region in which the electric field and the magnetic field are zero will still interact with the electromagnetic vector potential and scalar potential. The mechanism is the coupling of gauge fields with the complex phase of the charged particle's wave function.

4.8 Gauge Fields in Acoustics

When the gauge invariance is introduced into the acoustic field equations, the acoustics field of $\vec{v}$ and $\vec{T}$ can be written in terms of the gauge fields as:

$$\vec{v} = \text{curl}\,\vec{A} + \text{grad}\,\phi \tag{4.18}$$

and

$$\vec{T} = \nabla\phi - \frac{\partial \vec{A}}{\partial t} \tag{4.19}$$

Here, the concept of field strength will be first introduced into acoustics. Usually, in acoustics, the term acoustic intensity is used instead of field strength. Acoustic intensity is the square of the amplitude of the sound field. Field strength will give a more fundamental meaning.

For the electromagnetic waves, the field strength is given by

$$F_{\mu\nu} = \partial_\mu A_\nu - \partial_\nu A_\mu = \begin{bmatrix} 0 & -E_x/c & -E_y/c & -E_z/c \\ E_x/c & 0 & -B_z & B_y \\ -E_y/c & B_z & 0 & -B_x \\ E_z/c & -B_y & B_x & 0 \end{bmatrix} \tag{4.20}$$

For the acoustic waves, the electric field is analogous to the stress field and the magnetic field is analogous to the velocity field. Hence, the acoustic field strength will be given by

$$F_{\mu\nu} = \partial_\mu A_\nu - \partial_\nu A_\mu = \begin{bmatrix} 0 & -T_x/c & -T_y/c & -T_z/c \\ T_x/c & 0 & -V_z & V_y \\ T_y/c & V_z & 0 & -V_x \\ T_z/c & -V_y & V_x & \end{bmatrix} \tag{4.21}$$

References

1. Gan, W.S.: Gauge invariance approach to acoustic fields. In: Akiyama, I. (ed.) Acoustical Imaging, vol. 29, pp. 389–394. Springer, The Netherlands (2007)
2. Aharonov, Y., Bohm, D.: Significance of electromagnetic potentials in the quantum theory. Phys. Rev. **115**, 485 (1959)
3. Chambers, R.G.: Shift of an electron interference pattern by enclosed magnetic flux. Phys. Rev. Lett. **5**, 3 (1960)

Chapter 5
Covariant Derivative in Gauge Theory

Abstract First, covariance is explained. Then, the relation between covariant derivative and tensor analysis is described. The role of covariant derivative in local gauge invariance is given. The applications of gauge covariant derivative for sound propagation in continuous fluids and for sound propagation in solids are provided.

5.1 Covariance

Covariant derivative in gauge theory originated from the covariance used in the general theory of relativity. In the general theory of relativity, there is a condition of general covariance which states that all the equations of physics remain in the same form under the coordinate transformation for all forms of coordinates. The general theory of relativity is a gauge theory. The coordinates used here are the curvilinear coordinates which satisfy general covariance.

5.2 Covariant Derivative and Tensor Analysis

The gauge covariant derivative is commonly used in gauge transformation [1] to apply to the fields. This will enable simpler derivation of equations because it absorbs the external gauge field. The covariant derivative is an essential building block of gauge theory.

The form of the covariant derivative will be decided by the gauge transformation characteristics of the field. In general relativity, the partial derivative is replaced by the covariant derivative to satisfy the principle of general covariance. Also covariant derivative has to be written in tensor form to satisfy general covariance. The gauge covariant derivative is a partial derivative written in tensor form. It is a modification of the partial derivative to enable the equations to preserve their form and physical properties under gauge transformations.

W. S. Gan, *Gauge Invariance Approach to Acoustic Fields*,
https://doi.org/10.1007/978-981-13-8751-7_5

5.3 Covariant Derivative and Local Gauge Invariance

In local gauge invariance, the covariant derivative has to be introduced to replace the partial derivative. This will be in the following form:

$$D_t = \partial_t + G \tag{5.1}$$

where G is the additional gauge field to enable the local gauge invariance of the Lagrangian.

In mechanics, the Lagrangian can be expressed as:

$$\mathrm{L} = \mathrm{P.E.} - \mathrm{K.E.}$$

where P.E. = potential energy and K.E. = kinetic energy.

$$\mathrm{L} = \frac{1}{2}\left(\partial_{\mu\phi}\right)^T \partial_\mu \,\phi - \frac{1}{2} m^2 \phi^T \phi \tag{5.2}$$

where ϕ = vector fields and T^α matrices are generators of the SO(n) groups.

The Lagrangian is invariant under the global gauge transformation of:

$$\phi' = G\,\phi \tag{5.3}$$

where G = constant matrix.

The derivative of ϕ also follows global gauge transformation as:

$$(\partial_\mu \phi)' = G \partial_\mu \phi \tag{5.4}$$

For local gauge invariance of the Lagrangian, G will be the function of spacetime coordinates χ where $G = G(\chi)$.

$$\partial_\mu (G\phi) \neq G(\partial_\mu \phi) \tag{5.5}$$

That is, there is broken local gauge invariance in the Lagrangian.

To enable the local gauge invariance, a new derivative operator known as the gauge covariant derivative D_μ is introduced. Then

$$(D_\mu \phi)' = G\; D_\mu \phi \tag{5.6}$$

The covariant derivative is of the form

$$D_\phi = \partial_\mu - ig\,A_\mu \tag{5.7}$$

where g = coupling constant and $\overrightarrow{A_\mu}$ = vector potential, a gauge field.

The vector potential is an element of the Lie algebra and can be expressed as:

$$\overline{A_\mu} = \sum_\alpha A_\mu^\alpha T^\alpha \tag{5.8}$$

Local gauge invariance form of the Lagrangian will be:

$$L = \frac{1}{2}(D_\mu \phi)^T D_\mu \phi - \frac{1}{2} m^2 \phi^T \phi \tag{5.9}$$

5.4 Gauge Covariant Derivative for Sound Propagation in Continuous Fluids

Gauge transformation approach has been applied to fluid dynamics by Kambe [1]. Here, the fluid flow field is characterized by the translation symmetry. The fluid is interpreted as a material in motion characterized thermodynamically by mass density and entropy. Translation symmetry is also known as Galilean transformation. The Lagrangian of a system of point masses obeys Galilean symmetry known as global gauge invariance. Here for fluid flow, being a continuous field, in addition to global symmetry, the local gauge invariance of the Lagrangian is also required. If the fluid is in turbulence, rotational symmetry besides translational symmetry will also be required.

The gauge transformation in fluid flow will be useful for the treatment of gauge invariance of sound wave propagation in fluids. For the local gauge invariance of the fluid flow, Kambe [1] has derived the gauge field G as $\vec{v} \cdot \nabla$. Hence, the gauge covariant derivative will be:

$$D_t = \partial_t + \vec{v} \cdot \nabla \tag{5.10}$$

5.5 Gauge Covariant Derivative for Sound Propagation in Solids

For the application of gauge transformation to sound propagation in solids, the crystal is taken as an example. There will be global gauge invariance, in the form of Galilean symmetry as the crystals will be equivalent to point masses. In the case of local gauge invariance, there will be broken translation symmetry, giving rise to the phonon mode. The presence of phonons will interact with the electrons in the solids, giving rise to electron–phonon interaction and phonon–phonon interaction.

Reference

1. Kambe, T.: Gauge principle and variational formulation for ideal fluids with reference to translational symmetry. Fluid Dyn. Res. **39**, 98–120 (2007)

Chapter 6
Lie Groups

Abstract Group theory is introduced. The history of the Lie group is provided. This is followed by the description of the following classes of Lie groups commonly used: the unitary group of degree n, the special unitary group of degree n, the orthogonal group of degree n and special orthogonal group of degree n. Finally, the group properties of SO(n) with $n \geq 3$ are stated.

6.1 Group Theory

Group theory is the mathematics of symmetry. It is the study of symmetries. A group is the collection of symmetries of any object or equation. Group theory studies the algebraic structures known as groups. It is central to abstract algebras. It can help in the analysis of equations or objects which have symmetry properties. According to Richard Feynmann, all fundamental laws of physics have symmetry properties. The equations of physics are formulated and explained using the approach of symmetry. Symmetry is a form of invariance which does not change when subjected to a set of transformation. Group theory is the branch of mathematics which deals with the structure of groups and the mappings between groups. It is a powerful mathematical tool for analysing a physical system with symmetry properties. It is of great importance and wide-spread applications in physics, especially in particle physics and in condensed matter physics. It is said that particle physics cannot exist without group theory. It helps to predict the existence of several new elementary particles before their experimental discoveries. It has been used to determine the structure of crystals and provides a deep understanding of the theory of molecular vibration. It encodes the symmetry features of equations or objects and with a set of transformations that leave the equations or objects unchanged by performing successive transformations, for example, applying to a geometric object such as a circle as highly symmetric as being invariant under any rotation. Also, the function $x^2 + y^2 + z^2$ is invariant under any rearrangement of x, y and z and the trigonometric functions $\cos(t)$ and $\sin(t)$ are invariant when t is replaced by $t + 2\pi$. A group is the collection of symmetries of any object. Hence, every group is the symmetries of some equations or objects. Groups

W. S. Gan, *Gauge Invariance Approach to Acoustic Fields*,
https://doi.org/10.1007/978-981-13-8751-7_6

can be finite as well as infinite. The number of elements of a finite group is called the order of the group.

Symmetry of the equations of physics under various transformations gives rise to the conservation laws, such as the conservation of momentum, energy, electric charge. The conservation of energy arises from the physical laws remain unchanged in time. This is invariance under translation in time. The laws of physics also remain unchanged wherever we are in the universe. This invariance under translation in space leads to the law of the conservation of momentum. Also, the laws of physics under suitable rotation in space lead to the conservation of angular momentum. There is a theorem due to Emmy Noether which states that the symmetries of the physical systems give rise to conservation laws of physics.

The study and understanding of group theory are important in physics. This is because group theory describes the symmetries which the physical laws will obey. Besides this, group representation such as Lie groups point the way to new physical theories. Also, group theory plays an important role in the algebraic method for solving quantum mechanics problems. In quantum mechanics, the study of the group is related to the corresponding algebraic method. Today, the algebraic method is a subject of interest in various fields of physics. The algebraic method was first introduced in the subject of matrix mechanics in around 1925. The introduction of angular momentum in quantum mechanics intimately connected with the representation of the rotation group SO(3) associated with the rotational invariance of central potentials.

6.2 History of Lie Groups and Introduction

Lie group [1] is a branch of group theory. It is the most common form of gauge theory used in physics. Lie groups are the symmetry groups used in the standard model of particle physics. Lie group was named after the Norwegian mathematician Sophus Lie towards the end of the nineteenth century. He used this to study the symmetry in the solutions of differential equations. During his time, the techniques for solving the differential equations were to use several different tricks. One of them which was his idea was to use a skilled change of variables which would cause one of the variables to disappear from the equations. His idea worked. This was due to an underlying symmetry in the equations. This symmetry was now known as the Lie group. He founded the theories of continuous groups and their applications to the solutions of differential equations. He introduced Lie group to model the continuous symmetries of differential equations. Lie group's group operators are smooth and laid the foundation of the theory of continuous transformation. Lie group is a group of symmetries where the symmetries are continuous. An example of a continuous group of symmetries is a circle. One can rotate the circle in several steps of arbitrarily small increments and the circle will still look the same. Another example of continuous symmetry is rotational symmetry in three dimensions.

Besides being a smooth manifold obeying the symmetry property, Lie group has an additional condition that the group operators have to be differentiable. Lie groups combine group theory and analysis and are therefore the proper methods for describing the symmetries of equations and objects. The circle and the sphere are examples of smooth manifolds. The simplest cases of Lie groups are a one-dimensional real line under addition and the circle under rotation. In general, there are more complex group structures which will be described in the following sections.

The sphere has rotational symmetries in three dimensions. If one rotates the sphere by any angle in space, the sphere does not change in form. The collection of all these rotations will form the Lie group. The Lie group property means that if one rotates the sphere by an angle α, followed by angle β, there will be no change in the form of the sphere if one were to rotate the sphere in one attempt. There is an additional property that every rotation has an inverse, that is rotating in the opposite direction. This also demonstrates the concept of continuous symmetries of the Lie group.

The theory of Lie groups is to replace the global object by its local version which is its infinitesimal group or its Lie algebra [2]. Lie group represents the symmetry of a physical object or equation. The algebraic structure and the geometric structure must be compatible in a precise way. Hence, the representation of the Lie group or its Lie algebra is especially important. This is known as the representation theory which is used extensively in particle physics. Examples of groups whose representations are of importance are the rotation group SO(3) and the special unitary groups SU(2) and SU(3).

In the following sections, one will discuss more complicated Lie group structures besides circles.

6.3 Unitary Group of Degree n $U(n)$

The unitary group of degree n denoted by $U(n)$ is the group of $n \times n$ unitary matrices [3] with matrix multiplication as the group operation. The unitary group is a subgroup of the general linear group GL(n, C). $U(n)$ preserves the standard inner product as $C^{n.}$ For the simplest case of $n = 1$, group $U(1)$ corresponds to the circle group.This consists of the multiplication of all complex numbers with absolute value 1. The unitary group $U(n)$ represents a real Lie group of dimension n^2.

In mathematics, a complex square matrix U is unitary if its conjugate transpose $U*$ is also its inverse, that is,

$$U^*U = UU^* = 1 \tag{6.1}$$

where I = identity matrix.

Any unitary matrix A can be diagonalized by another unitary matrix S. A diagonal unitary matrix will have complex numbers of absolute value 1 on the main diagonal. This can be written as:

$$A = S\,\mathrm{diag}(e^{i\theta_1}, \ldots, e^{i\theta_n})S^{-1} \tag{6.2}$$

To illustrate the symmetry property of the unitary group, one will show that the unitary group can be defined in terms of quantity which are invariant after the unitary transformation:

Consider a complex transformation in two dimensions:

$$X' = Ax \tag{6.3}$$

Expressed in matrix form, this is:

$$\begin{pmatrix} x' \\ y' \end{pmatrix} = \begin{pmatrix} a & b \\ c & d \end{pmatrix}\begin{pmatrix} x \\ y \end{pmatrix} \tag{6.4}$$

a, b, c and d are complex. Here, there are eight free parameters.

$$\begin{aligned} /x'/^2 + /y'/^2 &= /ax + by/^2 + /cx + dy/^2 \\ &= ((ax + by)(a^*x^* + b^*y^*) + (cx + dy) + (c^*x^* + d^*y^*) \\ &= /x/^2 + /y/^2 \end{aligned} \tag{6.5}$$

This shows the invariance in the unitary transformation with these conditions on the matrix elements:

$$\begin{aligned} &/a/^2 + /c/^2 = 1 \quad /b/^2 + /d/^2 = 1, \\ &ab^* + cd^* = 0 \end{aligned}$$

6.4 Special Unitary Group of Degree *n* SU(*n*)

SU(n) stands for a special unitary group of degree n. It is the Lie group of $n \times n$ unitary matrices. The word "special" is used because of the determinant of the matrix being 1. Matrix multiplication is the group operation. It is a subgroup of the unitary group $U(n)$. It preserves the standard inner product as C^n. C stands for complex. SU(n) is a real Lie group. It is a real manifold with dimension n^2–1. It has simple Lie algebra and is a simple Lie group. It is compact and simply connected topologically.

SU(n) group has several applications in the standard model of particle physics, such as the SU(2) being used in the electroweak interaction and the SU(3) in the quantum chromodynamics. The simplest case of the SU(n) group is the SU(1) having only one element.

The Lie algebra of SU(n) consists of n $\times$ n skew-Hermitian matrices with trace zero. In linear algebra, a skew-Hermitian matrix is a square matrix with complex elements when its conjugate transpose is the negative of the original matrix. That is,

$$A^H = -A \tag{6.6}$$

The following is an example of SU(2) group. Since "special", the determinant of the unitary transformation is unity. The transformation matrix will have the form:

$$\begin{pmatrix} x' \\ y' \end{pmatrix} = \begin{pmatrix} a & b^* \\ b & -a \end{pmatrix} \begin{pmatrix} x \\ y \end{pmatrix} \tag{6.7}$$

where a is real. Also $/a/^2 + /b/^2 = 1$.

There are three free parameters correspond to three generators and SU(2) stands for the group of three matrices. In SU(2), the 2 refers to the size of the matrix. So, SU(2) is the group of 2 × 2 complex unitary matrices. The elements of SU(2) are all complex matrices of the form:

$$\mathrm{A} = \begin{pmatrix} z_1 & z_2 \\ z_3 & z_4 \end{pmatrix}$$

6.5 Orthogonal Group of Degree *n* *O*(*n*)

An orthogonal group is a Lie group of $n \times n$ orthogonal matrices and with matrix multiplication as the group operation. These matrices form a group because they are closed under multiplication and they can take the inverse. The orthogonal group is a Lie group because it is a group and a manifold. The definition of an orthogonal matrix is that it is a real matrix with its inverse equals to its transpose.

$$A^T := A^{-1} \tag{6.8}$$

Orthogonal matrix stands for a square matrix whose rows and columns are orthogonal unit vectors:

$$A^{\mathrm{T}}A = \mathrm{A}A^{\mathrm{T}} = I \text{ where } I = \text{identity matrix.} \tag{6.9}$$

What is identity matrix?

In linear algebra, an identity matrix of size n is a nxn square matrix with ones on the diagonal and zeros elsewhere. Denoting by I_n, one has:

$$I_1 = [1],\ I_2 = \begin{bmatrix} 1 & 0 \\ 0 & 1 \end{bmatrix},\ I_3 = \begin{bmatrix} 1 & 0 & 0 \\ 0 & 1 & 0 \\ 0 & 0 & 1 \end{bmatrix},$$

$$I_n = \begin{bmatrix} 1 & 0 & 0 \ldots\ldots\ldots\ldots & 0 \\ 0 \;\ldots & 1 \ldots\ldots 0 & & 0 \\ 0 & \ldots & & \\ \ldots & 0 & 0 & 0 \\ 0 & 0 & 0 & 0 \\ 0 & 0 & 0 & 1 \end{bmatrix}$$

What is invertible matrix?

In linear algebra, an n by n square matrix A is an invertible matrix if another n by n square matrix B such that $AB = BA = I_n$ where I_n is the n by n identity matrix and ordinary matrix multiplication is used. With such condition, then the matrix B is uniquely determined by A and is known as the inverse of A, denoted by A^{-1}

The following will explain the meaning of transpose of a matrix. The transpose of a matrix is another matrix whose rows are given by the columns of the original matrix. That is, the ith row and the jth column element of A^T will become the jth row and the ith column element of A:

$$\begin{pmatrix} 5 & 4 & 3 \\ 4 & 0 & 4 \\ 7 & 10 & 3 \end{pmatrix}^{\mathrm{T}} = \begin{pmatrix} 5 & 4 & 7 \\ 4 & 0 & 10 \\ 3 & 4 & 3 \end{pmatrix}$$

This is represented by the following equation:

$$\left[A^T\right]_{ij} = [A]_{ji} \tag{6.10}$$

When A is a $m \times n$ matrix, then A^T will be an $n \times m$ matrix.

An orthogonal group is also known as a rotation group. Its Lie algebra consists of the skew-symmetric $n \times m$ matrices with the Lie bracket [ ] given by the commutator. The determinant of an orthogonal group can be either 1 or −1.

An example of an orthogonal group is group O(2). This is defined as a group of 2×2 matrices with real elements and OO^{T} is the identity matrix. Here, the group action is two-dimensional and is rotational on the plane:

$$O(2) = \left\{\begin{bmatrix} \cos\theta & -\sin\theta \\ \sin\theta & \cos\theta \end{bmatrix}\right\} \mathrm{U} \left\{\begin{bmatrix} -\cos\theta & \sin\theta \\ \sin\theta & \cos\theta \end{bmatrix}\right\} \tag{6.11}$$

where θ is any real number in $[0, 2\pi]$.

These matrices have the quadratic form $x^2 + y^2$ and so represents the group orbits as circles $x^2 + y^2$.

6.6 Special Orthogonal Group of Degree *n* SO(*n*)

The special orthogonal group SO(n) is the subgroup of the elements of general orthogonal group with determinant 1. Common examples are SO(2) is a rotation group for the two-dimensional plane and the SO(3) is a rotation group for the three-dimensional space.

The following will give an illustration of an example of a SO(n) group. The special orthogonal group for $n = 2$ will be:

$$\mathrm{SO}(2) = \{A\varepsilon\, O(2) : \det A = 1\} \tag{6.12}$$

Here A εSO(2) with

$$A = \begin{pmatrix} \cos\theta & -\sin\theta \\ \sin\theta & \cos\theta \end{pmatrix}$$

Let $\begin{pmatrix} a\ b \\ c\ d \end{pmatrix}$ ε SO(2), then

$$\begin{pmatrix} a\ b \\ c\ d \end{pmatrix}\begin{pmatrix} a\ b \\ c\ d \end{pmatrix} = \begin{pmatrix} a^2 + c^2 & ab + cd \\ ab + cd & b^2 + d^2 \end{pmatrix} = \begin{pmatrix} 1\ 0 \\ 0\ 1 \end{pmatrix} \text{ and } \det \begin{pmatrix} a\ b \\ c\ d \end{pmatrix} = ad - bc = 1$$

This shows that SO(2) satisfies the following equations:

$$a^2 + c^2 = 1 \tag{6.13}$$

$$b^2 + d^2 = 1 \tag{6.14}$$

$$ad - bc = 1 \tag{6.15}$$

$$ab + cd = 0 \tag{6.16}$$

Equations (6.13) and (6.14) show that (a, c) and (b, d) lie on a circle. Thus, $a = \cos\alpha$, $c = \sin\alpha$, $b = \cos\beta$ and $d = \sin\beta$ for the angles α, β ε R. Inserting these substitutions in Eqs. (6.15) and (6.16),

One will have:

$$\cos\alpha \sin\beta - \cos\beta \sin\alpha = 1$$

$$\cos\alpha \cos\beta + \sin\alpha \sin\beta = 0$$

Using the angle sum trigonometric identities, the above equations can be simplified as:

$$\sin(\beta - \alpha) = 1$$

$$\cos(\beta - \alpha) = 0$$

The dimension of SO(2) is 1 because there is only one parameter, the rotation angle used to parameterize the group. The 2 refers to the size of the matrix. So, SO(2) is the group of 2×2 real orthogonal matrices. The numeral 2 refers solely to the dimension of the vector space in the definition of the group and not to other dimensions of representation.

Below we will show further group properties of SO(2). It is the rotation group in the two-dimensional plane. We need a representation for general elements of the group in order to have a better understanding of its group properties.

We will start by choosing an orthonormal basis for R^2, $\beta = \{(1, 0), (0, 1)\}$, where R stands for real. For the rotation by an arbitray angle θ, denoted by R_θ, the basics vector $(1, 0)$ will transform in the following manner:

$$R_\theta(1, 0) = (\cos\theta, \sin\theta) \tag{6.17}$$

Next one will rotate the other basis vector (0, 1) to obtain:

$$R_\theta(0, 1) = (-\sin\theta, \cos\theta) \tag{6.18}$$

Now, a matrix representation of R_θ can be constructed with respect to β using the transformed vectors as columns in the matrix:

$$\mathrm{A} = \begin{pmatrix} \cos\theta & -\sin\theta \\ \sin\theta & \cos\theta \end{pmatrix} \tag{6.19}$$

It is to be noted that $\det \mathrm{A} = \cos\theta^2 + \sin\theta^2 = +1$ which fulfils the requirement for the orthogonal group to be "special". Now,

$$\begin{aligned} \mathrm{A}A^T &= \begin{pmatrix} \cos\theta & -\sin\theta \\ \sin\theta & \cos\theta \end{pmatrix} \begin{pmatrix} \cos\theta & \sin\theta \\ -\sin\theta & \cos\theta \end{pmatrix} \\ &= \begin{pmatrix} \cos^2\theta + \sin^2\theta & \cos\theta\sin\theta - \cos\theta\sin\theta \\ \sin\theta\cos\theta - \cos\theta\sin\theta & \sin^2\theta + \cos^2\theta \end{pmatrix} = \begin{pmatrix} 1 & 0 \\ 0 & 1 \end{pmatrix} \end{aligned} \tag{6.20}$$

The above calculation shows that ordinary rotations in the two-dimensional plane that are contained in the SO(2) group.

What is an Abelian group?

In abstract algebra, an Abelian group is a group in which the group operation on two group elements does not depend on the order in which they are written. Therefore, an Abelian group is also known as a commutative group and obeys the axiom of commutability. SO(2) group is an example of an Abelian group and can be demonstrated as follows.

Let two elements of SO(2) be

$$\mathrm{A} = \begin{pmatrix} \cos\theta & -\sin\theta \\ \sin\theta & \cos\theta \end{pmatrix} \text{ and } \mathrm{B} = \begin{pmatrix} \cos\phi & -\sin\phi \\ \sin\phi & \cos\phi \end{pmatrix}$$

Then

$$\begin{aligned} \mathrm{AB} &= \begin{pmatrix} \cos\theta & -\sin\theta \\ \sin\theta & \cos\theta \end{pmatrix} \begin{pmatrix} \cos\phi & -\sin\phi \\ \sin\phi & \cos\phi \end{pmatrix} \\ &\begin{pmatrix} \cos\theta\cos\phi - \sin\theta\sin\phi & -\cos\theta\ sin\phi - \sin\theta\cos\phi \\ \sin\theta\cos\phi + \cos\theta\sin\phi & -\sin\theta\sin\phi + \cos\theta\cos\phi \end{pmatrix} \\ &= \begin{pmatrix} \cos(\theta+\phi) & -\sin(\theta+\phi) \\ \sin(\theta+\phi) & \cos(\theta+\phi) \end{pmatrix} \end{aligned} \tag{6.21}$$

If one swaps the elements A and B, then

$$\begin{aligned} \mathrm{BA} &= \begin{pmatrix} \cos\phi & -\sin\phi \\ \sin\phi & \cos\phi \end{pmatrix} \begin{pmatrix} \cos\theta & -\sin\theta \\ \sin\theta & \cos\theta \end{pmatrix} \\ &= \begin{pmatrix} \cos(\theta+\phi) & -\sin(\theta+\phi) \\ \sin(\theta+\phi) & \cos(\theta+\phi) \end{pmatrix} \end{aligned} \tag{6.22}$$

Hence, for A, B εSO(2), $AB = BA$. This proves that SO(2) is an Abelian group.

6.7 Group Properties of SO(n), with $n \geq 3$

This is an extension of the two-dimensional case of SO(2) to the three-dimensional rotation of the SO(3) group. The elements of the SO(3) group are represented by 3D. The determinants of the matrices will become more complicated as one moves to the higher order in n. Here, one can make use of the intuition that one has about rotations and matrices to make the work easier. Since one already knows what the elements of SO(2) look like, one can take one and insert it in a 3×3 matrix, such as:

$$\mathrm{A} = \begin{pmatrix} \cos\theta & -\sin\theta & 0 \\ \sin\theta & \cos\theta & 0 \\ 0 & 0 & 1 \end{pmatrix}, \ \mathrm{B} = \begin{pmatrix} 1 & 0 & 0 \\ 0 & \cos\theta & -\sin\theta \\ 0 & \sin\theta & \cos\theta \end{pmatrix} \tag{6.23}$$

The above matrices will produce $A^{\mathrm{T}} A = B^{\mathrm{T}} B = I$ and $\det A = \det B = 1$; hence, these matrices fulfil the requirements of SO(3). Besides this, they contain the block elements of SO(2). Hence, this technique of inserting matrices from SO(2) into higher dimensions can be used for constructing elements of SO(n).

What is a non-Abelian group?

SO(3) group is a non-Abelian group. This can be illustrated as follows:
Take the product AB from (6.23), and this will give

$$\mathrm{AB} = \begin{pmatrix} \cos\theta & -\sin\theta\cos\theta & sin^2\theta \\ \sin\theta & cos^2\theta & -\sin\theta\cos\theta \\ 0 & \sin\theta & \cos\theta \end{pmatrix}$$

Next, consider *BA* from (6.23); this gives

$$BA = \begin{pmatrix} \cos\theta & -\sin\theta & 0 \\ \sin\theta\cos\theta & cos^2\,\theta & -\sin\theta \\ sin^2\,\theta & \sin\theta\cos\theta & \cos\theta \end{pmatrix}$$

This shows that $AB \neq BA$ and hence, SO(3) is non-Abelian because it is not commutative. That is, rotating 90° along one axis followed by 90° along a different axis is different from rotating in the other way round. Both the continuous group and discrete group may be non-Abelian. In fact, most of the interesting Lie groups are non-Abelian and they play an important role in gauge theory. Furthermore, the elements of SO(3) may be embedded in SO(n) for $n > 3$. In fact, SO(2) is the only special orthogonal group that is Abelian because it describes rotation in the two-dimensional plane.

Non-Abelian group is sometimes called a noncommutative group.

The Lie algebra of the SO(3) group consists of a set of 3 × 3 skew-symmetric matrices. The generators of SO(3) correspond to the derivation of rotations around each of the standard axes, given by:

$$G_1 = \begin{pmatrix} 0 & 0 & 0 \\ 0 & 0 & -1 \\ 0 & 1 & 0 \end{pmatrix},\ G_2 = \begin{pmatrix} 0 & 0 & 1 \\ 0 & 0 & 0 \\ -1 & 0 & 0 \end{pmatrix},\ G_3 = \begin{pmatrix} 0 & -1 & 0 \\ 1 & 0 & 0 \\ 0 & 0 & 0 \end{pmatrix}$$

The Lie algebra of the SO(2) group corresponds to the set of 2 × 2 skew-symmetric matrices. SO(2) group has only one generator that corresponds to the derivation of two-dimensional rotation; it is given by:

$$G = \begin{pmatrix} 0 & -1 \\ 1 & 0 \end{pmatrix}$$

An element of SO(2) is any scalar multiple of the generator $\theta \,\varepsilon\, R$ and $\theta G \,\varepsilon$ SO(2)

SO(3) is the group of all rotations about the origin of three-dimensional Euclidean space R^3 under the operation of composition. A rotation about the origin is a transformation that preserves the origin, orientation (or handedness of space), the angles between vectors and Euclidean distance. A matrix will preserve or reverse orientation according to whether the determinant of the matrix is positive or negative. With the preservation of the Euclidean distance, it is isometric. Beside this, the rotation group has the manifold as its natural structure and possess smooth operations. Hence, it satisfies the conditions of a Lie group.

Every rotation maps an orthonormal basis of R^3 to another orthonormal basis.

Rotations are linear transformations of R^3. Therefore, they can be represented by matrices when the basis of R^3 is known. Once the orthogonal basis of R^3 is chosen, every rotation can be described by an orthogonal 3×3 basis. SO(3)is used in physics to describe the rotational symmetry and orientation of an object in space, giving rise to the elementary particle of integer spin.

Let R be a given rotation with respect to the standard basis, e_1, e_2, e_3 of R^3, the columns of R are given by $(R\ e_1,\ R\ e_2,\ R\ e_3)$. Since R preserves length and angle and the standard basis is orthonormal, the columns of R form another orthonormal basis. One can then express the orthonormality conditions as the following forms: $R^{\mathrm{T}}R = RR^{\mathrm{T}} = I$ where R^T denotes the transpose of R and I is the 3×3 identity matrix. The group of all 3×3 orthonormal matrices is denoted by O(3). For an orthogonal matrix R, with det R^{T} = det R implies $(\det R)^2 = 1$ which implies $R = \pm 1$. Here SO(3) is the special orthogonal group denoted SO(3) and is the subgroup of orthogonal matrices with determinat $+1$. This means that every rotation can be represented uniquely by an orthogonal matrix with determinant equals 1. Also, the composition of rotation corresponds to matrix multiplication, the rotation group is isomorphic to the special orthogonal group SO(3).

References

1. Hall, C.B.: Lie Groups, Lie Algebras, and representations: an elementary introduction. In: Graduate Texts in Mathematics, **222**, 2nd edn. Springer (2015)
2. Larson, R., Edwards, B., Falvo, D.: Elementary Linear Algebra, 5th edn. Houghton Mifflin, Boston, MA (2004)
3. Tapp, K.: Matrix Groups for Undergraduates, Student Mathematical Library, vol. 29. American Mathematical Society, Providence, RI (2005)

Chapter 7
Global Gauge Invariance

Abstract The meaning of gauge invariance is given. The U(1) group symmetry is introduced. Then, the gauge invariance approach to acoustics fields is provided.

7.1 What Is Gauge Invariance

Gauge invariance is a property of gauge theory. It is also known as gauge symmetry. In physics, the term gauge means any specific mathematical formalism used to regulate redundant degrees of freedom in the Lagrangian. So, gauge theory is a form of field theory in which the Lagrangian is invariant under local transformations. A gauge theory is a theory where the action is invariant under a continuous group symmetry that depends on spacetime. The continuous symmetry that depends on spacetime is called a space group. The transformations between gauges are known as gauge transformations form a symmetry group or the gauge group of the theory known as Lie group. There is the Lie algebra of group generators associated with any Lie group. For each generator, there is a corresponding field, a vector field called the gauge field. Gauge invariance is the invariance of the Lagrangian under local group transformations, and here, the gauge fields are included in the Lagrangian. There are both Abelian and non-Abelian Lie groups. If the symmetry group is commutable, the Lie group is an Abelian group. If the symmetry group is noncommutable, it is a non-Abelian group. The Maxwell equations of electromagnetism are an example of an Abelian group and the Yang–Mills theory of particle physics is a non-Abelian group.

Many important theories in physics are expressed in terms of Lagrangians that are invariant under certain symmetry transformation groups. If they are invariant under a transformation identically performed at every point within the spacetime in which the physical processes occur, then they are said to have a global symmetry or a global gauge invariance, That is, at any spacetime point, the transformation acts in the same way. That is, it is independent with respect to the space coordinates. For a local gauge invariance, the transformations are different at different points in the spacetime. In fact, the global gauge invariance is just a local gauge invariance in which the group parameters are fixed in spacetime. The Maxwell equations are an example of global

W. S. Gan, *Gauge Invariance Approach to Acoustic Fields*,
https://doi.org/10.1007/978-981-13-8751-7_7

gauge invariance and the Yang–Mills theory is an example of local gauge invariance. In fact, the gauge invariance in electromagnetism was recognized before the arrival of quantum mechanics. Gauge symmetry plays a very important role in quantum mechanics. Also, gauge invariance is the basis of electromagnetism.

So, there are two types of gauge invariance, the global gauge invariance and the local gauge invariance. In this chapter, only the global gauge invariance will be considered.

7.2 The U(1) Group Symmetry

Here, one will start with the wave function, instead of dealing with the Lagrangian as mentioned above. In the global gauge invariance, the phase change in the wave function is the same at all points in space. According to Noether's theorem, this will lead to the conservation of electronic charge in quantum electrodynamics. An example of global gauge invariance is the U(1) gauge symmetry in quantum electrodynamics. This global gauge symmetry arises is due to the fact that one is unable to measure the absolute phase of the electron wave function. U(1) group is the group of rotation around a fixed axis. It is an Abelian group. This explains the electromagnetic effect on the wave function of a charged electron.

The global gauge invariance for the quantum electrodynamic system can be represented by:

$$\Psi' = \Psi e^{i\theta} \tag{7.1}$$

where ψ = wave function before gauge transformation and ψ' = wave function after the U(1) group symmetry transformation.

In (7.1), the multiplication of ψ by $e^{i\theta}$ means a rotation of θ in the complex plane. Physically, this is the change of the phase of a plane wave by an angle θ. Here, a reference point is chosen arbitrarily for the measurement of the phase of plane waves. So, this is a global gauge invariance. If the phases of all plane waves are changed by the same amount, then this is equivalent to just moving the reference point. In fact, the global gauge invariance is a special case or a subset of the local gauge invariance in which the phase of each point is changed independently. This means that all observations are unchanged if one makes a global change of the phase of the wave function $\psi \to e^{i\theta}$.

In global gauge invariance $/\psi >$ denotes the same physical state as $e^{i\theta}\Psi >$. The U(1) rotation $/\psi > \to e^{i\theta}/\psi >$ on all states $/\psi >$ is known as the global gauge transformation.

7.3 Gauge Invariance Approach to Acoustic Fields

In 2007, Woon Siong Gan introduced gauge invariance approach to acoustic fields [1]. The above formulations which are for the electromagnetic fields can be applied to the acoustics fields. The concept of global gauge invariance and local gauge invariance also exists for acoustic fields.

Reference

1. Gan, W.S.: Gauge invariance approach to acoustic fields. In: Akiyama, I. (ed.) Acoustical Imaging, vol. 29, pp. 389–394. Springer, The Netherlands (2007)

Chapter 8
Local Gauge Invariance

Abstract First is the understanding of gauge fields. The next is the Lagrangian and the path integral approach to local gauge invariance. The introduction of covariant derivative into local gauge invariance is given followed by the local gauge invariance treatment in acoustic fields.

8.1 Understanding of Gauge Fields

To understand gauge invariance, one must start with the understanding of gauge fields. Under the quantum situation, in a local gauge invariance, each point in the spacetime will rotate independently. With local gauge invariance, each point will have a different phase that rotates the fields. Unlike global gauge invariance, where the phase change is the same at all points in space. The gauge invariance here is a form of rotational invariance. From Noether's theorem, the global gauge invariance leads to the conservation of charge in quantum electrodynamics. This is analogous to the translational invariance in classical physics which leads to the conservation of momentum. The local gauge invariance follows a non-Abelian group transformation. An example is the SU(2) group transformation. Here, one has to replace the partial derivative by the covariant derivative. In the local gauge invariance, the Lagrangian is invariant under certain Lie groups transformation. One can extend global gauge invariance into local gauge invariance by bringing in interaction between originally noninteracting fields.

8.2 The Lagrangian and the Path Integral Approach to Local Gauge Invariance

Applying action principle and path integral approach and consider a system of non-interacting real scalar fields with equal masses m. Using path integral, the action of the system will be given by:

W. S. Gan, *Gauge Invariance Approach to Acoustic Fields*,
https://doi.org/10.1007/978-981-13-8751-7_8

$$S = \int d^4 \times \sum_{i=1}^{n} \frac{1}{2}\partial_\mu \phi_i \partial^\mu \phi_i - \frac{1}{2}m^2\phi_i^2 \tag{8.1}$$

Introducing a vector of fields

$$\phi = (\phi_1, \phi_2 \ldots \phi_n)^T \tag{8.2}$$

one can write the Lagrangian as:

$$L = \frac{1}{2}(\partial_\mu \phi)^T \partial^\mu \phi - \frac{1}{2}m^2\phi^T \phi \tag{8.3}$$

where ∂_μ = Einstein notation for the partial derivative of ϕ in each of the four dimensions.

The Lagrangian is invariant under the global transformation of $\phi' = G\phi$ where G = constant matrix under the n by n orthogonal group $O(n)$. This means also

$$(\partial_{\mu\phi})' = \mathrm{G}\partial_\mu \phi \tag{8.4}$$

and

$$\partial_\mu (G\phi) = G(\partial_\mu \phi) \tag{8.5}$$

Extending the global gauge invariance to the local gauge invariance will require that the G matrices instead of being constants should be function of the spacetime coordinates χ. That is, $G = G(\chi)$ and

$$\partial_\mu (G\phi) \neq G(\partial_\mu \phi) \tag{8.6}$$

8.3 The Introduction of Covariant Derivative into Local Gauge Invariance

To enable the local gauge invariance, an additional term has to be added and the partial derivative ∂_μ has to be replaced by the covariant derivative D_μ.

Then,

$$(D_\mu \phi)' = G D_\mu \phi \tag{8.7}$$

and

$$D_\mu = \partial_\mu - \mathrm{ig} A_\mu \tag{8.8}$$

where g = coupling constant which defines the strength of an interaction.

So the gauge field $A(\chi)$ or the four-potential will transform as:

$$A'_{\mu} = \mathrm{G}\, A_{\mu} G^{-1} + \frac{i}{g}(\partial_{\mu} G)^{G-1} \quad (8.9)$$

A_{μ} is an element of the Lie algebra and can be expanded as $A_{\mu} = \sum_{\alpha} A_{\mu}^{\alpha} T^{\alpha}$.

There are as many gauge fields as there are generators of the Lie algebra.

The Lagrangian for global gauge invariance (8.3) will become:

$$L_{\mathrm{loc}} = \frac{1}{2}(D_{\mu}\phi)^{T} D_{\mu}\phi - \frac{1}{2}m^{2}\phi^{T}\phi \quad (8.10)$$

Pauli used the term gauge transformation of the first kind for global gauge invariance and gauge transformation of the second kind for local gauge invariance.

It can be shown that

$$\begin{aligned} L_{\mathrm{local}} - L_{\mathrm{global}} &= L_{\mathrm{interaction}} \\ &= i\frac{g}{2}\phi^{T} A_{\mu}^{T} \partial^{\mu}\phi + i\frac{g}{2}(\partial_{\mu}\phi)^{T} A^{\mu}\phi - \frac{g^{2}}{2}(A_{\mu}\phi)^{T} A_{\mu}\phi \end{aligned} \quad (8.11)$$

This interaction Lagrangian will take account of the interaction between the n scalar gauge fields or scalar potentials.

In local gauge invariance, one can view this additional interaction Lagrangian as the outcome of an additional gauge field to be added to make the whole mathematical formulation to be invariant when the rotation of each local point is being considered. The additional gauge field is communicating phase changes from one point to another. This gauge field is also associated with a force. In the case of quantum electrodynamics, this gauge field is the photon field, the carrier of the electromagnetic force.

The covariant derivative is also dealt with in a separate chapter of this book.

For local gauge invariance, the symmetry group is continuous such as the unitary group and the special unitary group. Local gauge invariance also gives rise to non-Abelian Lie groups and the phenomenon of spontaneously symmetry breaking (SSB).

8.4 Local Gauge Invariance Treatment in Acoustic Fields

So far the interpretation of local gauge invariance has been with the electromagnetic waves. In 2007, Woon Siong Gan introduced gauge invariance into acoustics fields [1]. Local gauge invariance can find applications in the applications to the treatment of acoustic multiple scattering such as to diffraction tomography and to the propagation of acoustic waves in solids and in fluids in condensed matter physics which will be dealt with in other chapters in this book.

Reference

1. Gan, W.S.: Gauge invariance approach to acoustic fields. In: Akiyama, I. (ed.) Acoustical Imaging, vol. 29, pp. 389–394. Springer, The Netherlands (2007)

Chapter 9
Gauge Fixing

Abstract The meaning and the procedure of gauge fixing are given. They are then illustrated by the examples of Maxwell's equations, the Coulomb gauge, and the Lorenz gauge.

In gauge theory, gauge fixing is a necessary procedure as one needs to choose the proper gauge. This will affect the subsequent derivation of the theories. This will involve the two basic gauge fields, the vector potential and the scalar potential and the deployment of them at the proper places. Gauge fixing will simplify the complexity of the equations. By fixing a gauge, the mathematical system, for instance, a set of partial differential equations will become simpler to analyse. The tractability of the resulting problem will depend on the choice of gauge one is going to fix. In the analysis of gauge theories, a key question is that one should decide how to fix a gauge and this will require the input of geometric ideas and intuition in the analysis.

In electromagnetism, the $U(1)$ symmetry provides invariance under the transformation of the four-potential A^{μ}:

$$A^{\mu\prime} = A^{\mu} + \partial^{\mu} \wedge (x) \tag{9.1}$$

where $\wedge(x) =$ a local function of spacetime coordinates.

Here, choosing a gauge amounts to choosing a particular four-potential A^{μ} the mathematical system. In fact, there are many useful gauges that one can fix. There is the Lorenz gauge: $\partial^{\mu} A_{\mu} = 0$ which is Lorentz invariant and hence transforms Maxwell's equations and the Yang–Mills equation into linear and nonlinear wave equations, respectively. Another important and useful gauge is the Coulomb gauge: $\nabla \cdot \vec{A} = 0$. The correct selection of a gauge is crucial in order to establish the basic properties of the underlying equation. It is possible that a given gauge field equation which is well-posed with one choice of gauge but ill-posed with another. Hence in order to use the standard partial differential equation method to solve the gauge field equation, it is necessary to first fix the gauge to a sufficient extent that this situation will not happen.

To illustrate the implication of gauge fixing, one will use the oldest gauge theory, the Maxwell's equations of electrodynamics. The gauge fields here are the magnetic

W. S. Gan, *Gauge Invariance Approach to Acoustic Fields*,
https://doi.org/10.1007/978-981-13-8751-7_9

vector potential $\vec{A}$ and the electric scalar potential ϕ, satisfying the following gauge fixing conditions:

$$\vec{E} = -\nabla\phi - \frac{\partial\vec{A}}{\partial t} \tag{9.2}$$

and

$$\vec{B} = \nabla \times \vec{A} \tag{9.3}$$

where $\vec{E}$ = electric field and $\vec{B}$ = magnetic field.

If one performs the following gauge transformations on $\vec{B}$ and $\vec{E}$

$$\overrightarrow{A'} \rightarrow \vec{A} + \nabla\phi \tag{9.4}$$

and

$$\phi' \rightarrow \phi - \frac{\partial\psi}{\partial t} \tag{9.5}$$

where ψ = a scalar function used to change the gauge called a gauge function.

Then, both $\vec{B}$ and $\vec{E}$ remain unchanged. That is $\overrightarrow{B'} = \vec{B}$ and $\overrightarrow{E'} = \vec{E}$.

The other two important and popular gauges are the Coulomb gauge and the Lorenz gauge.

(i) The Coulomb gauge is used in condensed matter physics and is defined by this gauge fixing condition:

$$\nabla \cdot \vec{A}(\vec{r}, t) = 0 \tag{9.6}$$

Then, the gauge fields can be represented by

$$\phi(\vec{r}, t) = \frac{1}{4\pi\varepsilon_0}\int\frac{\rho(\vec{r}, t)}{R}\mathrm{d}^3\overrightarrow{r'} \tag{9.7}$$

$$\vec{A}(\vec{r}, \mathrm{t}) = \nabla \times \int\frac{\rho(\vec{r}, \mathrm{t})}{4\pi R}\mathrm{d}^3\overrightarrow{r'} \tag{9.8}$$

where $\rho(\vec{r}, t)$ = electric charge density, $\vec{R} = \vec{r} - \overrightarrow{r'}$, $\vec{r}$ = any position vector in space, $\overrightarrow{r'}$ = a point in the charge or current distribution, and ε_0 = dielectric constant.

Also with the Coulomb gauge fixing condition, one can obtain the time-retarded electric current density $J(\vec{r}, \mathrm{t})$ as:

$$\vec{A}(\vec{r},t) = \frac{1}{4\pi\varepsilon_0}\nabla \times \int \left[\int_0^{R/c} \tau \frac{J(\vec{r};t-\tau)\times\vec{R}}{\vec{R}^3} \mathrm{d}\tau \right] \mathrm{d}^3\vec{r'} \tag{9.10}$$

(ii) The Lorenz gauge is given by:

$$\nabla . \vec{A} + \frac{1}{c^2}\frac{\partial\phi}{\partial t} = 0 \quad \text{in SI units} \tag{9.11}$$

$$\nabla . \vec{A} + \frac{1}{c}\frac{\partial\phi}{\partial t} = 0 \quad \text{in Gaussian units} \tag{9.12}$$

This may be written as

$$\partial_\mu A^\mu = 0 \tag{9.13}$$

where $A^\mu = \left[\frac{1}{c}\phi, \vec{A}\right]$ = electromagnetic four-potential and ∂^μ = 4-gradient.

The Lorenz gauge leads to the following inhomogeneous wave equations for the potentials:

$$\frac{1}{c^2}\frac{\partial^2}{\partial t^2}\phi - \nabla^2\phi = \frac{\rho}{\varepsilon_0} \tag{9.14}$$

$$\frac{1}{c^2}\frac{\partial^2}{\partial t^2}\vec{A} - \nabla^2\vec{A} = \mu_0 \mathrm{J} \tag{9.15}$$

Maxwell's equations in terms of the Lorenz gauge can be simplified to:

$$\partial_\mu \partial^\mu A^\nu = \mu_0 j^\nu \tag{9.16}$$

where $j^\nu = [c\rho, j]$ = 4-current.

The illustration of the above gauge fixing (9.2) to (9.5) is using the electromagnetic waves. This can be repeated for the acoustic fields equations as the gauge potentials ϕ and $\vec{A}$ also exist in acoustic fields [1] with the replacement of $\vec{E}$ and $\vec{B}$ by $\vec{T}$ and $\vec{v}$, respectively, where $\vec{T}$ = stress and $\vec{v}$ = sound particle velocity.

Reference

1. Gan, W.S.: Gauge invariance approach to acoustic fields. In: Akiyama, I. (ed.) Acoustical Imaging, vol. 29, pp. 389–394. Springer, Netherlands (2007)

Chapter 10
Noether's Theorem

Abstract The significance of Noether's theorem in gauge theory is given. Applications in physics are illustrated by a few examples. To apply Noether's theorem, the Lagrangian mechanics has to be used.

Emmy Noether's famous paper, Invariante Variationsprobleme [1], was published in Nachr. d. Konig. Gesellsch. d. Wiss. zu Gottingen, Math-phys. Klasse in 1918 [1]. In this paper, Noether proves two different theorems. The first theorem deals with "global" symmetries (generated by finite Lie groups) and states that these symmetries lead to conserved charges. The second theorem applies to local gauge symmetries (infinite-dimensional Lie groups), containing arbitrary functions of spacetime (like Einstein's theory of gravity), and shows that these gauge symmetries inevitably lead to relations among the equations of motion (e.o.m. onwards).

Emmy Noether's first theorem states that the differentiable symmetry of the action of a physical system has a corresponding conservation law. She proved this theorem in 1915 and subsequently published in 1918 [1]. This comes under calculus of variations and Lagrangian mechanics and Hamiltonian mechanics. It has a limitation only to systems that can be modelled with a Lagrangian alone. That is whenever there is a continuous symmetry of Lagrangian, there is an associated conservation law. Symmetry here means that any transformation of the generalized coordinates q, of the associated velocities $\dot{q}$, and of the time variable t will leave the Lagrangian unaffected. Continuous symmetry means a symmetry with a continuous constant parameter, typically infinitesimal and measures how far from the identity the transformation is bringing us.

In action principle or the principle of least action, the action of a physical system is the integral over time of a Lagrangian function. The symmetries here cover continuous and smooth symmetries over the physical space. If the Lagrangian of a physical system is invariant or symmetric under continuous rotations, then from Noether's theorem the angular momentum of the system will be conserved under the laws of motion. If the Lagrangian is invariant or symmetry regardless of place or time, then it has translational symmetry in space and time, and Noether's theorem states that there will be conservation in linear momentum and energy within the system respectively. If the Lagrangian of the system is invariant or symmetry upon

W. S. Gan, *Gauge Invariance Approach to Acoustic Fields*,
https://doi.org/10.1007/978-981-13-8751-7_10

spatial or temporal reflection, then it has reflection symmetry and time reversal symmetry, respectively. Then, there will be conservation of parity and entropy according to Noether's theorem.

Noether's theorem has important applications in physics. It enables the determination of conserved quantities from the symmetries of the Lagrangian. Conversely, it enables one to derive the whole classes of Lagrangians with invariance to describe a physical system. For instance, with a proposed theory with conservation of a certain quantity, one can derive the forms of Lagrangian that will conserve the quantity through continuous symmetry. This will provide the criteria to understand the conditions and on the correctness of the new theory.

Noether's theorem originated from the Euler–Lagrangian equation of mechanics which gives the following equation of motion:

$$\frac{\mathrm{d}}{\mathrm{d}t}\left(\frac{\partial L}{\partial \dot{q}}\right) = \frac{\partial L}{\partial q} \tag{10.1}$$

where

L = Lagrangian and
q = generalized coordinates.

This equation implies that if the right-hand side of (10.1) is zero and L is symmetric over q, then the left-hand side is also zero meaning and the rate of change of the quantity in parentheses is also zero. So this is a conserved quantity.

Noether formulated this concept as Noether's theorem in 1915. As a simple example to illustrate the application of the Euler–Lagrangian equation (10.1), one will take L = Kinetic energy of a moving particle as $L = (1/2)\,m\,\dot{x}^2$. Also here, L depends on $\dot{x}$, not on x, so $\partial L/\partial x = 0$, that is the momentum is conserved.

The above example can be generalized to a system with n coordinates $q_1, q_2, q_3, \ldots q_n$. Then, the Lagrangian will become a function of n variables and their first derivatives with respect to time. The total differential of L with respect to all the n coordinates will be:

$$\mathrm{d}L = \sum_{i=1}^{n} \frac{\partial L}{\partial q_i}\mathrm{d}q_i \tag{10.2}$$

Multiplying each of the Euler equations by the respective differential $\mathrm{d}q_i$ and summing, one has:

$$\frac{\mathrm{d}}{\mathrm{d}t}\left[\sum_{i=1}^{n} \frac{\partial L}{\partial \dot{q}_l}\right]\mathrm{d}q_i = \mathrm{d}L \tag{10.3}$$

This shows that there is also a conserved quantity for the case of combination of differentials for the n coordinates.

It is well known that all physical laws have the symmetry of displacements in time. Here, the Lagrangian is not an explicit function of time and the total derivative of the Lagrangian is given as:

$$\frac{\mathrm{d}L}{\mathrm{d}t} = \sum_{i=1}^{n} \left[\frac{\partial L}{\partial q_i} \frac{\mathrm{d}q_i}{\mathrm{d}t} + \frac{\partial L}{\partial \dot{q}_i} \frac{\mathrm{d}\dot{q}_i}{\mathrm{d}t} \right] \tag{10.4}$$

The Euler–Lagrangian equations for the system enable us to substitute for the partials of L with respect to q_i on the right-hand side. Now expressing $\mathrm{d}q_i/\mathrm{d}t$ in dot notation, one has:

$$\frac{\mathrm{d}L}{\mathrm{d}t} - \sum_{i=1}^{n} \left\{ \dot{q}_i \frac{\mathrm{d}}{\mathrm{d}t} \left[\frac{\partial L}{\partial \dot{q}_i} \right] + \frac{\partial L}{\partial \dot{q}_i} \frac{\mathrm{d}\dot{q}_i}{\mathrm{d}t} \right\} = 0 \tag{10.5}$$

This is the derivative with respect to time of a single product and so can be rewritten as:

$$\frac{\mathrm{d}}{\mathrm{d}t} \left[L - \sum_{i=1}^{n} \dot{q}_i \frac{\partial L}{\partial \dot{q}_i} \right] = 0 \tag{10.6}$$

This means that the quantity within the square bracket is a constant.

This can be illustrated by the following example of a mass–spring system of three masses, m_1, m_2, and m_3. The spring constant between the first and second mass is k_{12}, and the spring constant between mass two and three is k_{23}. The displacements of the three masses are x_1, x_2, and x_3, respectively.

The Lagrangian of the system is given by

$$L = T - V = \frac{1}{2} m_1 \dot{x}_1^2 + \frac{1}{2} m_2 \dot{x}_2^2 + \frac{1}{2} m_3 \dot{x}_3^2 - \left[\frac{k_{12}}{2} (x_2 - x_1)^2 + \frac{k_{23}}{2} (x_3 - x_2)^2 \right] \tag{10.7}$$

So we have

$$\sum_{i=1}^{3} \dot{x}_i \frac{\partial L}{\partial \dot{x}_i} = m_1 \dot{x}_1^2 + m_2 \dot{x}_2^2 + m_3 \dot{x}_3^2 = 2T \tag{10.8}$$

Since $L = T - V$, time symmetry will produce a conserved quantity as $L - 2T = T - V - 2T = -(T + V)$ which is the total energy of the system. This is another outcome of Noether's theorem.

The above formulation is based on Lagrangian mechanics and is applicable to the acoustic fields which are mechanical in nature.

Reference

1. Noether, E.: Invariant variation problems. Gott. Nachr. **1918**, 235–257 (1918) [Transp. Theory Statist. Phys. **1**, 186 (1971)]

Chapter 11
Spontaneous Symmetry Breaking and Phonon as the Goldstone Mode

Abstract First is the Introduction to spontaneous symmetry breaking. Continuous symmetry is explained followed by phonon as a Goldstone mode and then Goldstone's theorem.

11.1 Introduction to Spontaneous Symmetry Breaking

Spontaneous symmetry breaking is also known as conditional symmetry. It means that the overall system obeys symmetry and only broken takes place at the ground state of the quantum system. This can be illustrated by the Mexican hat potential after Goldstone [1].

Mexican hat potential [1]

Here, one considers a symmetric upward dome with a trough circling the bottom. With a ball at the peak of the dome, the system is symmetric with respect to a rotation around the centre axis. When the ball rolls down the dome into the trough, which is at the ground state of the system or at the point of lowest energy, there is spontaneous symmetry breaking. Afterward the ball has come to rest at some fixed point on the perimeter. The dome and the ball preserve their individual symmetry, but the whole system undergoes a spontaneous symmetry breaking (Fig. 11.1).

This can be described by a scalar field theory. One starts with the Lagrangian of the scalar field $\boldsymbol{\phi}$. The Lagrangian of the system can be written as:

$$
\begin{aligned}
L &= \text{kinetic energy} - \text{potential energy} \\
&= \partial^{\mu}\boldsymbol{\phi}\partial_{\mu}\,\boldsymbol{\phi} - V(\boldsymbol{\phi})
\end{aligned}
\quad (11.1)
$$

The symmetry breaking is triggered by V($\boldsymbol{\phi}$). Goldstone [1] has given an example of this potential function, the "Mexican hat" potential:

$$
V(\boldsymbol{\phi}) = -10|\phi|^2 + |\phi|^4 \quad (11.2)
$$

W. S. Gan, *Gauge Invariance Approach to Acoustic Fields*,
https://doi.org/10.1007/978-981-13-8751-7_11

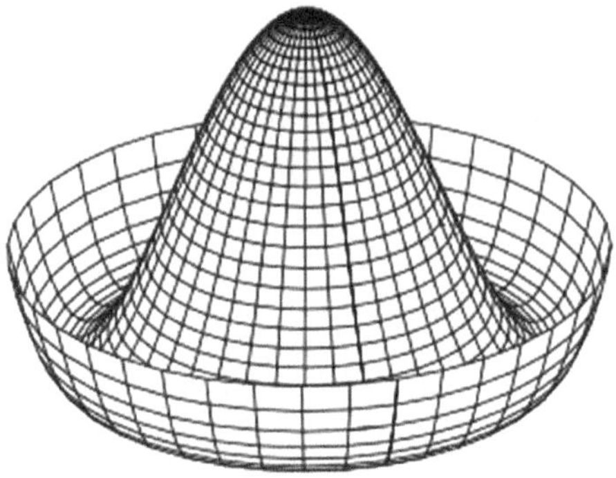

Fig. 11.1 Graph of Goldstone's "Mexican hat" potential function V($\boldsymbol{\phi}$) [1]

giving an infinite number of possible minima or vacuum states with value $\boldsymbol{\phi} = \sqrt{5}\, e^{i\theta}$. Θ can have real values between 0 and 2π. For $\boldsymbol{\phi} = 0$, the system has an unstable ground state or vacuum state which has a U(1) symmetry. With a choice of θ, the system will fall into a specific-state vacuum state, the spontaneous symmetry breaking will occur. The spontaneous symmetry breaking gives rise to the Goldstone mode [1], which runs around the circle at the minima of this potential. This means any other values of θ would also have exactly the same energy.

11.2 Continuous Symmetry—Phonon as a Goldstone Mode

Spontaneously breaking of a continuous symmetry is inevitably accompanied by gapless Goldstone [1] modes associated with slow long-wavelength fluctuations of the order parameter. Gapless means that those modes do not require any energy to excite. For example, vibrational modes in a crystal, known as phonons, are associated with slow density fluctuations of the crystal's atoms. For symmetry-breaking states, whose order parameter is not a conserved quantity, Goldstone modes are typically massless and propagate at a constant velocity.

Goldstone modes [1] are present in any system with a broken continuous symmetry. There are no Goldstone modes when a discrete symmetry is broken because it is impossible to produce a slowly varying rotation from one state to an equivalent one. Phonons are an example of Goldstone modes corresponding to the breaking of rotation and translational symmetries by a crystal structure.

Goldstone bosons are massless degrees of freedom that are consequences of the breaking of the continuous symmetry of the ground state. The Goldstone particles or modes become decoupled in the low energy limit and possess the following properties:

a. They can be represented by fields that are rotational scalars meaning that these particles or modes are spinless bosons.

b. There is one Goldstone mode or particle for each broken symmetry generator.

Spontaneous breaking of continuous symmetry is accompanied by massless or gapless Goldstone modes associated with slow long-wavelength fluctuations of the order parameter. Gapless means that these modes do not require any energy to excite. An example is the vibrational modes of a crystal known as phonons are associated with slow density fluctuations of the crystal's atoms. For symmetry-breaking states, the Goldstone modes are massless and propagate at a constant velocity such as phonons.

Phonon as Goldstone mode is a strong evidence of the symmetry property of the acoustic field. Phonon is a gauge field and similar to that the photon is also a gauge field for the case of electromagnetic wave. When sound wave propagates in a continuous medium, there is translational symmetry. However, when sound wave propagates in a discrete medium, like crystal structure, then there is spontaneous breakdown of translational symmetry giving rise to the phonon mode. The spontaneous breakdown of translational symmetry was discovered by Nambu [2] in the BCS theory of superconductivity and further explained by Goldstone [1]. Phonons are classified under bosons. Bosons and fermions are the two classes of particles of nature. The name boson was coined by Paul Dirac to commemorate Satyendra Nath Bose, Indian professor of physics at the University of Calcutta and University of Dhaka who developed with Albert Einstein the Bose–Einstein statistics of statistical mechanics which provides the characteristics of elementary particles. Boson is a force carrier, and it functions as the glue holding the matter together. Fermions, on the other hand, are particles that make up matter. Examples are electron, protons, neutrons, mesons, etc. Besides phonons, other examples of bosons are photons, gluons, W and Z bosons, the four force-carrying gauge bosons of the standard model, the recently discovered Higgs bosons and the hypothetical graviton of quantum gravity. Boson has the important quantum property that there is no restriction to the number of them occupying the same quantum state. Fermions, on the other hand, have the limitation that no two identical fermions can occupy the same quantum state. Also, bosons have integer spins and fermions have half-integer spins. However, phonons are spinless bosons.

11.3 Goldstone's Theorem

Goldstone's theorem states that if the Lagrangian or Hamiltonian or free energy is finitely ranged, and if a continuous symmetry is spontaneously broken, then massless or gapless excitations will be generated associated with the local reorientation of the broken symmetry. These massless excitations are known as Goldstone bosons.

Goldstone's theorem applies to a continuous symmetry which is spontaneously broken. The broken symmetry due to the ground state is not invariant under the action of the corresponding charges. It also gives rise to new massless scalar particles being generated. This scalar massless particle is known as the Nambu–Goldstone boson. Each generator of the broken symmetry will produce one Nambu–Goldstone boson. The Nambu–Goldstone mode is also a long-wavelength fluctuation of the corresponding order parameter.

One-dimensional lattice breaks continuous translational symmetry into discrete translational symmetry which generates one Goldstone boson or longitudinal phonon. In systems which break continuous translational symmetry, will also necessarily break rotational symmetry. This is not true the other way round. Examples of such systems are regular crystals, lamellar solids, etc.

For the case of translational symmetry, the gapless point of the phonon mode is a uniform translation of the lattice. This requires zero energy One can represent that phonon is gapless by expressing the energy $E(k)$ of a phonon of lattice momentum k and take the limit:

$$\lim_{k \to 0} E(k) = 0 \tag{11.3}$$

In fluids, the spontaneously broken Galilean symmetry produces the longitudinal phonon as the Goldstone boson. In solids, the spontaneously broken Galilean, translational symmetry produces the longitudinal phonons and the spontaneously broken rotational symmetry produces the transverse phonon. The longitudinal and the transverse phonons are the Goldstone bosons.

References

1. Goldstone, J., Salam, A., Weinberg, S.: Broken symmetries. Phys. Rev. **127**, 965–970 (1962)
2. Nambu, Y., Jona-Lasinio, G.: Dynamical model of elementary particles based on an analogy with superconductivity, I. Phys. Rev. **122**, 345 (1961)

Chapter 12
Time Reversal Acoustics and Superresolution

Abstract First, introduction is followed by the theory of time reversal acoustics with the following coverage: time reversal acoustics and superresolution. The following applications of time reversal acoustics are given to medical ultrasound imaging and to ultrasonic nondestructive testing. The application to nondestructive testing has the following coverage: theory of time reversal acoustics for liquid–solid interface, experimental implementation of the time reversal mirror for nondestructive testing work, incoherent summation, time reversal of signals coming from a speckle noise zone and the iterative technique. Then, there is also the application of time reversal acoustics to landmine or buried object detection with the following coverage: introduction, theory, experimental procedure, experimental set-up, Wiener filter and experimental results. Next, there is the application of time reversal acoustics to nonlinear acoustics. The application of time reversal acoustics to nonlinear acoustics, nonlinear acoustic imaging, and the experimental set-up to analyse the nonlinear time reversal acoustics waves.

12.1 Introduction

The new field of time reversal acoustics was pioneered by Fink in 1992 [1]. It is based on the time reversal invariance of the acoustic wave equation. It has the characteristics of working well in heterogeneous media where multiple scatterings take place instead of in ordered or homogeneous media. Hence, it enables sharp focusing in heterogeneous media. Since most solid media are inhomogeneous in nature, this technique has great potential in acoustical imaging and even shows the possibility of superresolution, that is, defeating the diffraction limit.

12.2 Theory of Time Reversal Acoustics

Time reversal acoustics is based in the principle of time reversal invariance of the acoustic wave equation in lossless media [1–4]. It means that, if $\varphi(\vec{r}, t)$ is an acoustic field and thus is a solution of the wave equation, then $\varphi(\vec{r}, -t)$ is another solution

W. S. Gan, *Gauge Invariance Approach to Acoustic Fields*,
https://doi.org/10.1007/978-981-13-8751-7_12

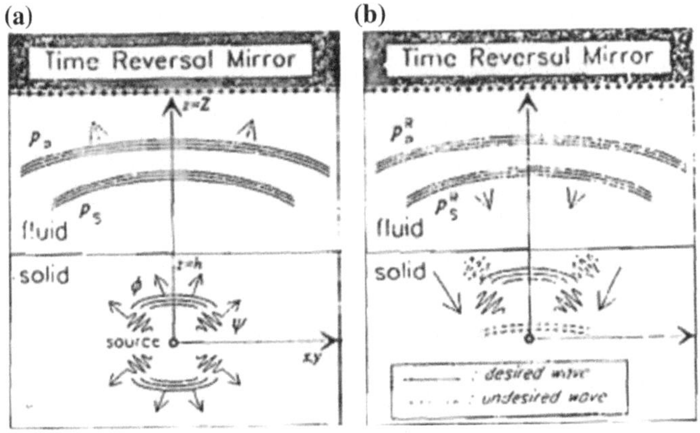

Fig. 12.1 Time reversal behaves as a two-step process. **a** Emission of a short pulse of longitudinal and transverse waves by an active source in the solid, yielding two pressure wavefronts in the fluid; recording by the TRM; **b** re-emission of the time-reversed fields into the fluid by the TRM; backpropagation of the two wavefronts in the fluid, yielding four wavefronts in the solid; the desired waves arrive simultaneously at the initial source location, the undesired ones arrive before and afterwards (from Draeger et al. [5])

and thus a possible acoustic field. In particular, if $\varphi(\vec{r}, t)$ is a wave diverging from a point, $\varphi(\vec{r}, -t)$ must be focused on this same location. This gives rise to the novel idea of focusing procedure in two steps: step one, an acoustic source is installed and its emitted field is measured by means of a closed receiving surface around the medium. We suppose that each point of this surface is able to record the wave as a function of time and then to re-emit in step two in order to generate the time reversal solution. This time-reversed wave is backpropagated through the medium and finally focuses on the locations of the initial source.

However, the concept of closed time-reversed cavity (CTRC) is difficult to realize [3] and is usually replaced by a time-reversed medium (TRM) of finite size which shows of performance comparable to the cavity in spite of the loss of information [4]. We will consider the case of solid–fluid interface which is applicable to situations of nondestructive testing and medical imaging.

Draeger et al. [5] consider a point-like source of elastic waves located inside a solid half space at the origin $x = y = z = 0$ (Fig. 12.1a). The plane solid–fluid interface is at $z = h > 0$ and the TRM is located in the fluid at $z = Z > H$. The source emits a short pulse of longitudinal and transverse waves (or P and S waves) which are partially terminated at the interface into the fluid and thus both converted into pressure waves. They consider that all waves emitted or reflected into the negative z directions are lost. SH waves, in particular, are totally reflected, and therefore, it is impossible to apply the time-reversed process to this polarization of transverse waves. This results in a limitation of the TRM device. That is why the P and SV wave components yield two wavefronts in the fluid.

The TRM records them both and is able to distinguish between the two wavefronts if their interval time is sufficiently distinct. In this case, they can choose to backpropagate only the wavefronts corresponding to the P waves or the SV waves or both. When the backpropagating wavefronts arrive at the fluid–solid interface, each of them creates two wavefronts in the solid, one corresponding to the original type of waves, thus wanted, and one of the other type of wave, thus unwanted (Fig. 12.1b). Also the unwanted SV wave created by the returned P wavefront and the unwanted P wave due to the returned SV wavefront yield a low-level noise, and they are not focused and arrive at different times. On the contrary, the two wanted wavefronts focus at the same place, that is, the location of the initial source which now remain passive.

Their derivations here are for the P and SV waves propagating in the positive Z direction [5]. They describe the displacement vector $\vec{u}$ of the incident elastic field as a function of the scalar and vector potentials ϕ and ψ of the P and SV waves propagation in the positive z direction [5]:

$$\vec{u}(x, y, z, t) = \nabla\phi(x, y, z, t) + \nabla x D \nabla x(0, 0, \psi(x, y, z, t)) \tag{12.1}$$

The initial field can be for a point-like source or extended source. The only condition is the existence of the 2D Fourier transform over x and y in the frequency domain of the potential $\Phi(k_x, k_y, z, \omega)$ and $\Psi(k_x, k_y, z, \omega)$: ϕ is the scalar potential and is given rise to the longitudinal wave field

$$\phi(x, y, z, t) = \frac{1}{2\pi} \int \mathrm{d}\omega\, \tilde{\phi}(x, y, z, \omega) xp(-j\omega t) \tag{12.2}$$

and

$$\tilde{\phi}(x, y, z, \omega) = \left(\frac{1}{2\pi}\right)^2 \iint \mathrm{d}k_x \mathrm{d}k_y \tilde{\Phi}(k_x, k_y, z, \omega) \exp(J(k_x x + k_y y)) FT_{2D}^{-1}[\Phi] \tag{12.3}$$

Referring to the acoustic wave Eq. (13.1), the dependence on z can be written explicitly as

$$\tilde{\Phi}(k_x, k_y, z, \omega) = \tilde{\Phi}(k_x, k_y, z = 0, \omega) \exp(j v_a(k_r) z) \tag{12.4}$$

where v_a is a function of $k_r = \sqrt{k_x^2 + k_y^2}$ and the longitudinal wavespeed α and is defined by

$$v_\alpha = \begin{cases} \sqrt{\omega^2/\alpha^2 - k_r^2}, & \text{if } k_r \le \frac{\omega}{\alpha} \\ j\sqrt{k_r^2 - \omega^2/\alpha^2}, & \text{if } k_r > \frac{\omega}{\alpha} \end{cases} \tag{12.5}$$

If υ_α is real, i.e., if $k_r \leq \frac{\omega}{\alpha}$, it can be considered as the z component k_z of the wave vector $\vec{k} = (k_x, k_y, k_z)$ of a propagative wave.

If υ_α is imaginary, the wave is called evanescent wave and is nonpropagative. Its amplitude decreases exponentially with the depth in the z direction.

Ψ is the vector potential, and it gives rise to the shear wave.

Following same procedure as for the longitudinal wave, one obtains:

$$\tilde{\Psi}(k_x, k_y, z, \omega) = \tilde{\Psi}(k_x, k_y, z = 0, \omega)\exp(j\upsilon_\beta(k_r)z) \tag{12.6}$$

Each wave type in the solid generates at the interface a wavefront in the fluid. Describing the transmitted sound wave by its pressure field P, they can split it formally into the past created by the P wave P_p and the one from the SV wave P_s:

$$P(x, y, z, t) = P_\text{p}(x, y, z, t) + P_\text{s}(x, y, z, t) \tag{12.7}$$

where

$$\tilde{P}_\text{p}(k_x, k_y, z, \omega) = \tilde{\Phi}(k_x, k_y, z = 0, \omega)T_{ps}\exp(j\upsilon_\alpha h)\exp(j\upsilon_c(z - h)) \tag{12.8}$$

and

$$\tilde{P}_\text{s}(k_x, k_y, z, w) = \tilde{\psi}(k_x, k_y, z = 0, \omega)T_{sf}\exp(j\upsilon_\beta h)\exp(j\upsilon_c(z - h)) \tag{12.9}$$

and T_{pf}, T_{sf} = transmission coefficients.

The derivation of the transmission coefficients is as follows:

The incident plane sound waves are on a solid–fluid or fluid–solid interface. For simplicity, the interface is at $z = 0$ and $k_y = 0$; $k_x = k_r$. Three cases are considered: (1) incident P wave, (2) incident SV wave. Referring to Eq. (12.1), the expression of the trial wave fields is

$$\begin{aligned} \text{Case (1)}\ \tilde{\phi} &= \exp(j(k_x x + \upsilon_\alpha z)) + R_{pp}\exp(j(k_x x - \upsilon_\alpha z)) \\ \tilde{\psi} &= R_{ps}\exp(j(k_x x - \upsilon_\beta z)) \\ \tilde{P} &= T_{pf}\exp(j(k_x x + \upsilon_c z)) \end{aligned}$$

where R_{pp}, R_{ps} = reflection coefficients.

$$\begin{aligned} \text{Case (2)}\ \tilde{\psi} &= \exp(j(k_x x + \upsilon_\beta z)) + R_{ss}\exp(j(k_x x - \upsilon_\beta z)) \\ \tilde{\phi} &= R_{sp}\exp(j(k_x x - \upsilon_\alpha z)) \\ \tilde{P} &= T_{sf}\exp(j(k_x x + \upsilon_c z)) \end{aligned}$$

The fields are related to each other by boundary conditions at the interface $z = 0$ [8, 9]:

(1) The z component of the displacement u_z is continuous:

$$\tilde{u}_z = \frac{\partial}{\partial z}\tilde{\phi} - \frac{\partial^2}{\partial x^2}\tilde{\psi} = \frac{1}{\omega^2 \rho}\frac{\partial}{\partial z}\tilde{p}$$

(2) The vertical traction is continuous [7]:

$$\tilde{T}_2 = \lambda\nabla^2\tilde{\phi} + 2\mu\frac{\partial^2}{\partial z^2}\tilde{\phi} - 2\mu\frac{\partial^3}{\partial x^2 \partial z}\tilde{\Psi} = -\tilde{p}$$

where λ and μ are the Lame's coefficients and related to the wavespeeds by $\lambda + 2\mu = \rho_s\alpha^2$ and $\mu = \rho_s\beta^2$.

(3) The horizontal traction T_x vanishes: [5]

$$\tilde{T}_x = 2\mu\frac{\partial^2}{\partial x \partial z}\tilde{\phi} - \mu\frac{\partial}{\partial x}\left(\frac{\partial^2}{\partial x^2}\tilde{\psi} - \frac{\partial^2}{\partial z^2}\tilde{\psi}\right) = 0$$

Eliminating the reflection coefficients for each case described above, the transmission coefficients are obtained as:

$$T_{pf} = \frac{2}{N}\rho_s\rho\omega^4\left(\omega^2 - 2\beta^2 k_r^2\right)v_\alpha \tag{12.10}$$

$$T_{ss} = \frac{4(-jk_r)}{N}\rho_s\rho\beta^2 w^4 k_r v_\alpha v_\beta \tag{12.11}$$

$$T_{fp} = \frac{2}{N}\left(\omega^2 - 2\beta^2 k_r^2\right)v_c$$

$$T_{fs} = \frac{-4}{N(-jk_r)}\beta^2 k_r v_\alpha v_c \tag{12.12}$$

where $N = 4\rho_s\beta^4 k_r^4 v_c + 4\rho_s\beta^4 k_r^2 v_\alpha v_\beta v_c - 4\rho_s\beta^2\omega^2 k_r^2 v_c + \rho_s\omega^4 v_c - \rho\omega^4 v_\alpha$

At $z = 2$, the TRM records the arriving field. Here, they consider the simpler case of phase conjugation in the frequency domain. They assume that the mirror is infinitely large, i.e., it measures and emits at each point in the whole xy-plane at $z = Z$. They omit the aperture function to simplify the forward mathematics. From Eqs. (12.8) and (12.9), they obtain for the time-reversed wavefronts:

$$\tilde{P}_p^R(k_x, k_y, z, \omega) = \tilde{\Phi}^*(k_x, k_y, z = 0, \omega)T_{pf}^* \exp(-jv_\alpha^* h)\exp(-jv_c^*(z - h))\exp(-jv_c(z - Z)) \tag{12.13}$$

and

$$\tilde{P}_s^R(k_x, k_y, z, \omega) = \tilde{\Psi}^*(k_x, k_y, z = 0, \omega)T_{sf}^* \exp\left(-jv_\beta^* h\right)\exp(-jv_c^*(z - h))\exp(-jv_c(z - Z)) \tag{12.14}$$

Each of them generates two wavefronts in the solid. Hence, they obtain two desired wave fields corresponding to the initial type of wave,

$$\tilde{\Phi}_p^R(k_x, k_y, z, \omega) = \tilde{\Phi}^*(k_x, k_y, z = 0, \omega)\exp(-jv_\alpha z)T_{pf}^* T_{fp}\exp(-2\mathrm{Im}t v_\alpha h)\exp(-2\mathrm{Im}t v_c(z-h)) \tag{12.15}$$

and

$$\tilde{\Phi}_s^R(k_x, k_y, z, \omega) = \tilde{\Psi}^*(k_x, k_y, z = 0, \omega)\exp(-jv_\beta z)T_{sf}^* T_{fs}\exp(-2\mathrm{Im}v_\beta h)\exp(-2\mathrm{Im}v_c(z-h)) \tag{12.16}$$

and two undesired wavefronts, one SV wave created by a returned P wave and vice versa:

$$\tilde{\Phi}_s^R(k_x, k_y, z, \omega) = \tilde{\Psi}^*(k_x, k_y, z = 0, \omega)\exp(-jv_\alpha z)T_{sf}^* T_{fp}\exp(\mathrm{j}(v_\alpha)\exp(-2\mathrm{Im}v_c(z-h)) \tag{12.17}$$

$$\tilde{\Phi}_p^R(k_x, k_y, z, \omega) = \tilde{\Psi}^*(k_x, k_y, z = 0, \omega)\exp(-jv_\beta z)T_{pf}^* T_{fs}\exp(j(v_\beta - v_\alpha^*)h)\exp(-2\mathrm{Im}v_c(z-h)) \tag{12.18}$$

With $v_\alpha = v_\alpha^* + 2j\mathrm{Im}v_\alpha$ and $FT_{2D}[\Phi^*\exp(-jv_\alpha^*)] = \tilde{\psi}^*(\mathrm{x}, \mathrm{y}, \mathrm{z}, \omega)$, the returned and desired P wave of Eq. (5.2.15) can be written in the xy-space as a convolution:

$$\tilde{\phi}_p^R(x, y, z, \omega) = \left(\frac{1}{2\pi}\right)^2 \tilde{\phi}^*(x, y, z, \omega) \underset{x,y}{*} FT_{2D}[T_{pf}^* T_{fp}\exp(-2\mathrm{Im}v_\alpha(h-z))\exp(-2\mathrm{Im}v_c(Z-h))] \tag{12.19}$$

The first term $\tilde{\phi}^*(x, y, z, \omega)$, corresponds exactly to the time-reversed P field $\phi(x, y, z, -t)$, we are interested in. But the quality of the reversed wave decreases by losses due to interface and propagation.

In the same procedure, the returned, desired SV wave can be written as:

$$\tilde{\Psi}_s^R(x, y, z, \omega) = \left(\frac{1}{2\pi}\right)^2 \tilde{\Psi}^*(x, y, z, \omega) \underset{x,y}{*} FT_{2D}[T_{sf}^* T_{fs}\exp(-2\mathrm{Im}v_\beta(h-z))\exp(-2\mathrm{Im}v_c(z-h))] \tag{12.20}$$

$\tilde{\Psi}^*(x, y, z, \omega)$ corresponds to the time-reversed SV field $\psi(x, y, z, -t)$. This means in particular that, if the TRM reverses both fields together, they will focus at the same time at the same place, i.e. the initial source location. This is the proof of the time reversal mirror capability of spatial and temral recompression.

The undesired wavefronts do not arrive at the same time as the desired ones at the origin. The time-reversed wavefront corresponding to the P wave generates at the interface two wavefronts in the solid. A desired one of P wave type and an undesired one of SV wave type which propagate more slowly than the first one and hence accumulate later at the initial source position in the same way, the time-reversed wavefront corresponds to the SV wave generates an undesired P wavefront in the solid which arrives later at the initial source positions. In the same way, the time-reversed wavefront corresponds to the SV wave generates an undesired P wavefront in the solid which arrives sooner than the desired one. So they find the main property of the undesired wave. They arrive at the wrong time, they are not focused and they are of low amplitude. The TRM can reverse the P and SV waves but cannot reverse the SH polarization. The reversed P and SV waves arrive at the same time, focused on the initial source location with a focal spot width corresponding approximately to their central wavelength. Thus, the slower transverse waves are better focused.

12.2.1 Time Reversal Acoustics and Superresolution

In time reversal acoustics, a signal is recorded by an array transducers, time reversed and then propagates back through the medium and refocuses approximately on the source that emitted it. The refocusing is approximate because of the finite size of the aperture of the array of transducers (receivers and transmitters) which is only a certain portion of the 3D time reversal cavity. It is often small compared to the propagation distance, so only a small part of the advancing wave is captured and time reversed. In a homogeneous medium, the refocusing resolution of the time-reversed signal is limited by diffraction. However, when the medium has random inhomogeneities, the refocusing effect is better and the resolution of the refocused signal can in some circumstances beat the diffraction limit. This is superresolution. In homogeneous media, the spatial resolution of the time-reversed signals is limited by diffraction and is inversely proportional to the aperture size and proportional to the wavelength times the propagation distance. Time-reversed signals propagate backwards through the time-independent medium and go through all the multiple scattering, reflection and refraction that they underwent in the forward direction. That is why refocusing occurs.

If the medium is randomly inhomogeneous, the focusing equation of the back-propagated signal can be better than the resolution in the homogeneous case. This is referred to as superresolution. The random inhomogeneities produce multipathing and the TRM appears to have an aperture that is larger than its physical size, an effective aperture $a_e > a$. This means that the recompressed pulse is narrower than in the homogeneous medium and we have superresolution with a spatial scale of order $\lambda L/a_e$ where L is distance from the source to the TRM and a is the size of the TRM. This phenomenon was observed in underwater acoustics experiments (Dowling and Jackson [6]; Hodgkiss et al. [7]; Kuperman et al. [8]) as well as in the ultrasound regime (Derode et al. [9]; Fink [10, 11]).

Practical examples of this illustration of superresolution have been demonstrated with computer simulation by Blomgren et al. [12] for underwater acoustical imaging. They presented a detailed analytical and numerical study of how multipathing in random media enhances resolution in time reversal acoustics. That is how superresolution arises in random media. They have shown that when the propagation distance is large compared to the wavelength and the correlation length of the inhomogeneities and the time reversal minor is small, there is an exact expression for the effective size of the TRM, its effective aperture valid in both in the time and frequency domain. Multipathing makes the effective size of the TRM much larger than its physical size.

Lehman and Devaney [13] have demonstrated superresolution applied to seismic imaging. They have applied their theory to the case where the transmitted and receiver sensor arrays need not be coincident and for cases where the background medium can be nonreciprocal. Their theory is based on the singular value decomposition of the generalized multistatic data matrix of the sensor system rather than the standard eigenvector/eigenvalue decomposition of the time-reversed matrix as was employed in other treatments of time reversal imaging. They derived a generalized multiple signal classification (MUSIC) algorithm that allows superresolution imaging of both well-resolved and nonwell-resolved point targets from arbitrary sensor array geometry. Their time reversal MUSIC algorithm is tested and validated in two computer simulations of offset vertical seismic profiling where the sensor sources are aligned along the earth's surface and the receiver array is aligned along a subsurface borehole. Their results demonstrated the high-contrast, high-resolution imaging capability of their new algorithm combination when compared with classical backpropagation or field focusing.

12.3 Application of Time Reversal Acoustics to Medical Ultrasound Imaging

The beauty of time reversal acoustics (TRA) in the application to medical ultrasound imaging is that the technology works even more precisely in a heterogeneous medium where there are lots of ultrasound-distorting obstacles. One of the challenges of imaging the human body or targeting tumours or gall stones nonsurgically is that human body is inhomogeneous tissues, fat and bone varying in density, so most ultrasound signals become scattered and distorted. Difficult and challenging environments like the human body in fact improve the focusing of ultrasound to a specific location, sharpening the focus and enhancing precision.

Another advantage of TRA in medical applications is that systems work so rapidly and refocus so quickly that movement of the body presents almost no problem.

The time reversal mirror (TRM) which consists of array of transducers that convert sound waves into electrical signals. Then a computer reverses their order and the transducers transfer electrical signals back into sound and target the reverse sound waves back in the direction from which they came. The process will keep on repeating until the required sharp focusing is achieved. This is in fact an iterative procedure.

An example of the medical application of TRD is the research carried out by Dr. Sutin [14] originally from Russia, now Senior Scientist at Davidson Laboratory, part of the Charles V. Schaefer School of Engineering at Stevens Institute of Technology. His research on TRA is supported by the National Institute of Health. The projects involved are:

(1) A "virtual finger" that could focus on an area inside the body much more precisely than any other known method. One of the challenges of imaging the human body or targeting tumours or gall stones nonsurgically is that the body is inhomogeneous and time reversal technology is especially suitable for focusing in inhomogeneous medium.

(2) Application of time reversal acoustics (TRA) to nonlinear imaging. Such imaging would involve several TRA focusing systems. The interaction of the crossing beam produces nonlinear effects and this allows three-dimensional, nonlinear acoustic images of an object inside a human body. TRA together with nonlinear imaging will enhance the focusing of sound waves in the inhomogeneous human body.
(3) Another application of a TRA/nonlinear technique is the use of ultrasound to measure the blood pressure inside a certain point or chamber within the heart. To make this work, harmless tiny capsules (ultrasound contact agents) would be introduced in the blood stream. They would react differently during the heart's different pressure conditions and their reactions would be measured by sound waves aimed from different angles and returned to a time reversal mirror. Variations in harmonics levels resonating from the capsules would be correlated to ambient pressure. The precision of the TRA system would allow highly accurate focusing in one area of the heart. Meanwhile, the nonlinear acoustic technique would give a diagnostician information about pressure changes as the heart pumps.

A different application would involve the ability to nonsurgically check for cracks in older mechanical heart values, a valuable tool for warding off difficulties in some heart patterns. The detection of internal cracks and flaws in materials is a standard use of nonlinear acoustic technique. This application would simply employ it inside the human body with the benefit of TRA focusing.

Dr. Sutin's blood pressure measurement involving time reversal acoustics has been supported by Steven Institute of Technology as part of the Technogenesis initiatives. Technogenesis is Stevens' unique environment for education and research, in which students, faculty and partners in academia, government and industry, jointly nurture new technologies and companies from laboratory innovation to marketplace implementation.

12.4 Application of Time Reversal Acoustics to Ultrasonic Nondestructive Testing

The time reversal mirror method for ultrasonic nondestructive testing is a novel and completely different approach to focusing on a defect beneath plane or curved surfaces. It is based on the concept of time reversal of ultrasonic fields and taken into account both the phase and modulus information coming from the defect. This technique is self-adaptive and requires only the presence of a target in the solid sample. In highly scattering media, it is shown that the time reversal process allows a new approach to speckle noise reduction. It is able to compensate for the distortions induced by liquid–solid interfaces of different geometries and to detect small defects in a noisy background.

Time reversal method is especially useful to detect small defects inside curved surfaces. Ultrasonic nondestructive testing needs large focusing apertures in order to detect small defects in solid media. Currently two approaches, both scanned immersion techniques, have been extensively studied in order to obtain focused beams through curved liquid–solid interfaces. Both techniques require a priori knowledge of the geometry of the interface.

In the first technique, the beam focusing is achieved with one or several transducers whose geometry is matched to the liquid–solid interface and to the desired focal point. In this technique, each transducer has a front face designed to equalize all the propagation times between the transducer surface and the desired focal point in the solid. However, due to the curved surface, these transducers are in focus for only one point in the solid (with a limited depth of focus). Industrial inspection of thick samples thus requires many different transducers.

The second technique uses a multi-element transducer to generate a focal spot in the acoustic beam at any specified angle and range. The 1D and 2D transducer arrays are connected to a set of electronics delay lines whose values are matched to produce focusing [15–17]. The optimal delay, calculated using Snell's law, compensates the propagation time variations between the different elements and the desired focal point. The focusing and beam steering ability result from the interference pattern produced by the delayed acoustic pulses.

The limitation of these two techniques is that they are based on an exact a priori knowledge of the interface geometry and require highly precise positioning of the transducer. As the transducer aperture becomes larger, the precision of the positioning needs to be higher and such a precision is not always available in current inspection. They also assume that the velocity of sound is known and constant in each propagating medium.

Time reversal acoustic is a novel and completely different approach for the focusing on defects beneath curved surfaces. It is based on the principle of time reversal invariance of the acoustic wave equation which is also known as the reciprocity theorem. The technique is self-adaptive and only needs the presence of a defect in the sample of interest. Hence, 1D and 2D transducer arrays are used. Neither a priori knowledge of the interface geometry nor knowledge of sound velocity in the propagating medium is required.

In the time reversal process, we take advantage of the properties of the piezoelectric transducers, i.e. their transmit and receive capabilities, their linearity and the capability for instantaneous measurement of the temporal pressure waveforms. The pressure field $p(\vec{r}, t)$ reflected by a defect is detected with a set of transducer elements located at positions r_i and is digitized and stored during a time interval T. The detected pressure fields are then resynthesized and transmitted by the same transducers in a reversal temporal chronology. This is equivalent to the transmission of $p(\vec{r}, T - t)$. Each transducer of the array is connected to its own electronic circuitry that consists of a receiving amplifier, an A/D converter, a storage memory and most importantly a programmable transmitter capable of synthesizing a time-reversed version of the stored signals.

Such a time reversal mirror (TRM) can converge a divergent wave issuing from a defect into a converging wave focusing on it. The TRM produces a real image of the defect at the position of the initial source. It is a self-focusing technique that compensates geometrical distortions of the array structure as well as those resulting from the propagation through liquid–solid interfaces. It can be achieved in real time with simple signal processing.

Another very attractive feature of time reversal technique is its speckle noise reduction capability. In highly scattering media, the detection of small defects is usually difficult due to the speckle noise generated by the heterogeneous structure. The TRM has the capability of distinguishing between the speckle noise and the reflected signal from the defect. If the speckle noise comes from a microstructure whose scale is less than the wavelength, the time reversal process cannot refocus on the speckle noise sources due to the loss of information during the propagation.

12.4.1 Theory of Time Reversal Acoustics for Liquid–Solid Interface [18]

Chakroun et al. [18] consider the environment of ultrasonic nondestructive testing as the immersion type with liquid–solid interface. The liquid could be gel or water. The basic theory starts with the acoustic wave equation in a lossless medium.

For sound propagation in a homogeneous fluid, with constant velocity c, the acoustic wave equation is

$$\frac{1}{c^2}\frac{\partial^2 p}{\partial t^2} = \nabla p \tag{12.21}$$

where

$p(\vec{r}, t)$ acoustic pressure field.

For sound propagation in an isotropic solid, the acoustic wave equation is

$$\rho_s \frac{\partial^2 u}{\partial t^2} = (\lambda + 2\mu)(\mathrm{grad}(\mathrm{div}(\vec{u})) - \mu(\mathrm{curl}(\mathrm{curl}\,\vec{u})) \tag{12.22}$$

where

ρ_s density of the solid and
λ, μ Lame's coefficients

Both Eqs. (12.21) and (12.22) contain only a second-order time-derivative operator. Thus, if $p(\vec{r}, t)$ and $u(\vec{r}, t)$ are solutions of (12.21) and (12.22), then $p(\vec{r}, -t)$ and $u(\vec{r}, -t)$ are also solutions of the same equations. Moreover, $p(\vec{r}, t)$ in the fluid and $u(\vec{r}, t)$ in the solid are linked by stress and strain continuity at the fluid–solid interface. This determines the unique solution $p(\vec{r}, t)$ for the pressure field in the

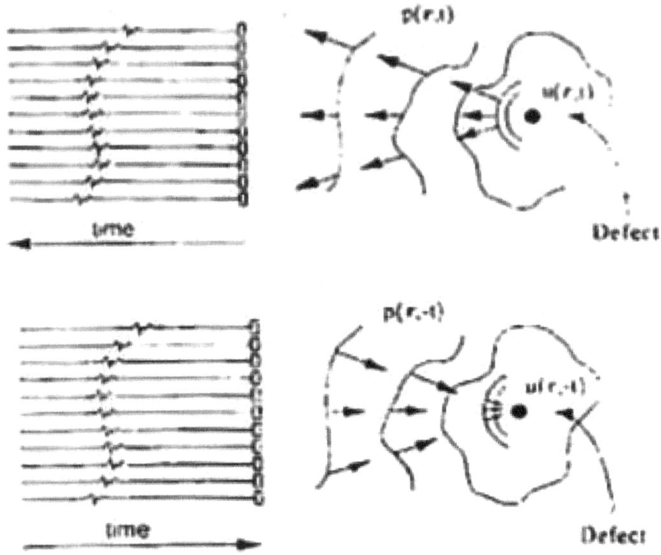

Fig. 12.2 Time reversal through liquid–solid interface (from Chakroun et al. [18])

fluid. This property can be used to achieve optimal focusing of a pulsed wave as a point-like reflector located in a solid sample immersed in fluid medium.

Let $u(\vec{r}, t)$ represents the acoustic displacement in the solid and $p(\vec{r}, t)$ the resulting pressure field in the fluid that propagates from a single-point source located in a solid. The optimal way to focus at this source consists of a time reversal of the pressure field in the whole 3D volume, generating $p(\vec{r}, -t)$ in the fluid and thus $u(\vec{r}, -t)$ in the solid (Fig. 12.2).

Using Huygens' principle, time reversal process can be reduced from a 3D volume to a 2D closed surface, resulting in the concept of a closed time reversal cavity [3]. Since a closed cavity is difficult to realize experimentally, this can be comprised by considering a portion of the cavity, such as a 2D piezoelectric array located in the fluid medium, in front of the solid sample. The TRM here work as follows. First, a pulsed wave is transmitted to the solid sample from some transducers of the array. The pressure field $p(\vec{r}, t)$ $(1 < i < N)$ scattered by a point-like target in the solid sample is then detected with the N elements of the array $\vec{r}$, digitalized and recorded within a time interval T. Finally, the pressure field is retransmitted by the same transducers in a reversed temporal chronology (last in, first out). This is equivalent to the transmission of $p(\vec{r}, T - t)$. Such a mirror approximately realizes the backpropagation of the field to its initial source and then focuses on the target in the solid.

12.4.2 Experimental Implementation of the TRM for Nondestructive Testing Works

Four steps are required for the experimental implementation of the time reversal process for nondestructive testing (NDT) works.

(1) Transmit Step No. 1: The first incident pulsed wave is transmitted from the liquid towards the solid by one or more elements of the array. In this first transmission, the array sends unfocused acoustic energy into the material.
(2) Receive Step No. 1: The echoes coming from the block are measured by the same 2D array on the N transducers. For each transducer k, the corresponding discrete signal $R_{k,1}[m]$ is recorded in a storage memory. If a defect is present in the illuminated volume, it behaves as an acoustic source and reflects a small amount of the energy transmitted in the previous step.
(3) Transmit Step No. 2: During this step, we choose the origin and temporal length of the signals to be time reversal. This is achieved through the definition of temporal window identical for all the transducers, each window corresponds to a given depth of inspection in the material. The depth of inspection is known by measuring the transmit time of the acoustic pulse as in conventional ultrasonic inspection. For each element k, the windowed signal $W_{k,1}[m]$ is time reversed and stored in the corresponding transmit memory. The new transmit signals, E_2, are used to transmit a second wave from the array towards the block.
(4) Receive Step No. 2: The new echoes coming from the sample R_2 are recorded. If the time reversal windows, previously selected W_1, contain information from a defect, the resulting time-reversed wave refocuses naturally on it and the signals W_2 show a high-level amplitude. The defect is now strongly amplified, it is cavity detected.

This can be illustrated by the following practical example: ultrasonic inspection of hard-alpha grains in titanium alloys through plane interfaces. This is an important effort to improve ultrasonic inspection of titanium alloys. In commercial titanium alloy, two kinds of phases of grain occur: alpha grain and beta grain. In jet engine components, both alpha and beta grains are present. During the elaboration process of titanium, metallurgical inclusion effects such as hard-alpha can appear. A hard-alpha inclusion is a localized region of alpha phase grains which have a substantially higher hardness and brittleness than the surrounding material. If not detected, hard-alpha inclusion can become crack initiation sites and lead to component failure. The detection sensitivity of this kind of defect is limited because titanium is an acoustically noisy medium. The strong ultrasonic speckle noise is induced by the polycrystalline microstructure. A second limitation comes from the characteristics of the hard-alpha: this defect has a low reflectivity due to a small acoustic impedance mismatch and has irregular and unknown shape.

Here, the time reversal process is compared for signals coming from two different zones of interest from a plane titanium sample. One zone contains a hard-alpha defect embedded in the titanium microstructure. The other contains only titanium microstructure. The experiment is carried out with a 2D array of transducers.

The time reversal NDT experiment to detect hard-alpha was done in the laboratory of *M*, Fink [19]. The followings are the four steps needed to focus ideally on the hard-alpha. The first incident wave is transmitted by the central element of the 2D array. Figure 12.3a shows the recorded data in grey level for reception 1 for the 121 transducer elements of the TRM. Data are presented in B-Scan mode where the horizontal axis represents time (equivalent to depth) and the vertical axis represents the transducer number. They correspond to the logarithmic envelope of the echographic signals received on the 121 elements of the array. The bottom line in Fig. 12.4a is the signal from element No. 1, the second line from element No. 2 and so on. From these data, we can see the echoes coming from the front and back faces of the titanium block. Between these high-amplitude echoes, we note the titanium speckle noise induced by the microstructure. This reflected sound results in a defect signal which is superimposed upon the grain noise background. Here, the defect signal cannot be

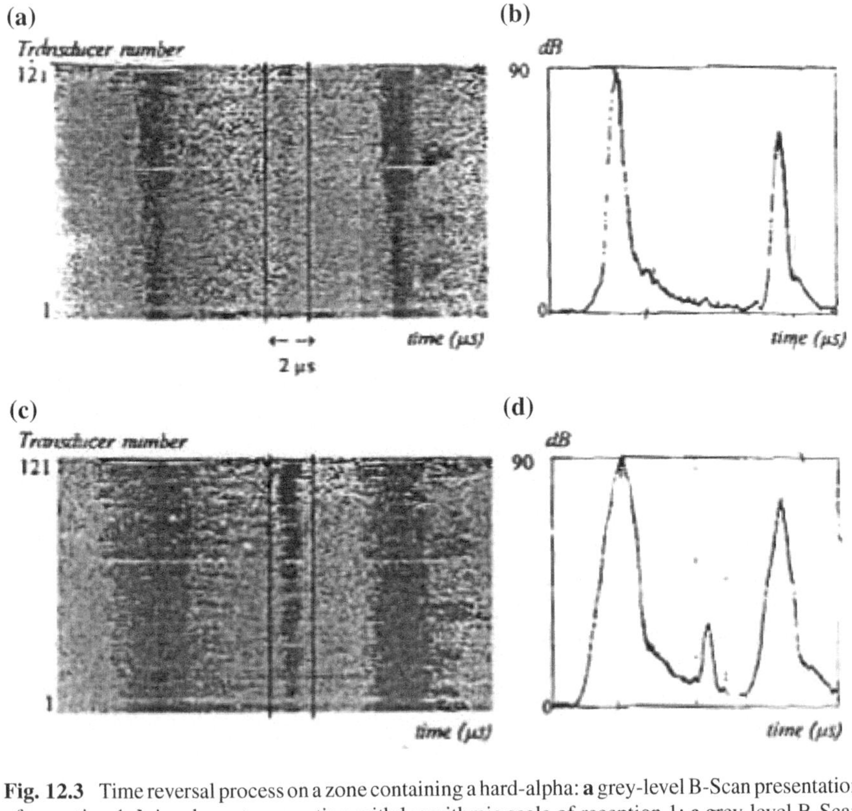

Fig. 12.3 Time reversal process on a zone containing a hard-alpha: **a** grey-level B-Scan presentation of reception 1; **b** incoherent summation with logarithmic scale of reception 1; **c** grey-level B-Scan presentation of reception 2; **d** incoherent summation with logarithmic scale of reception (from Chakroun et al. [18])

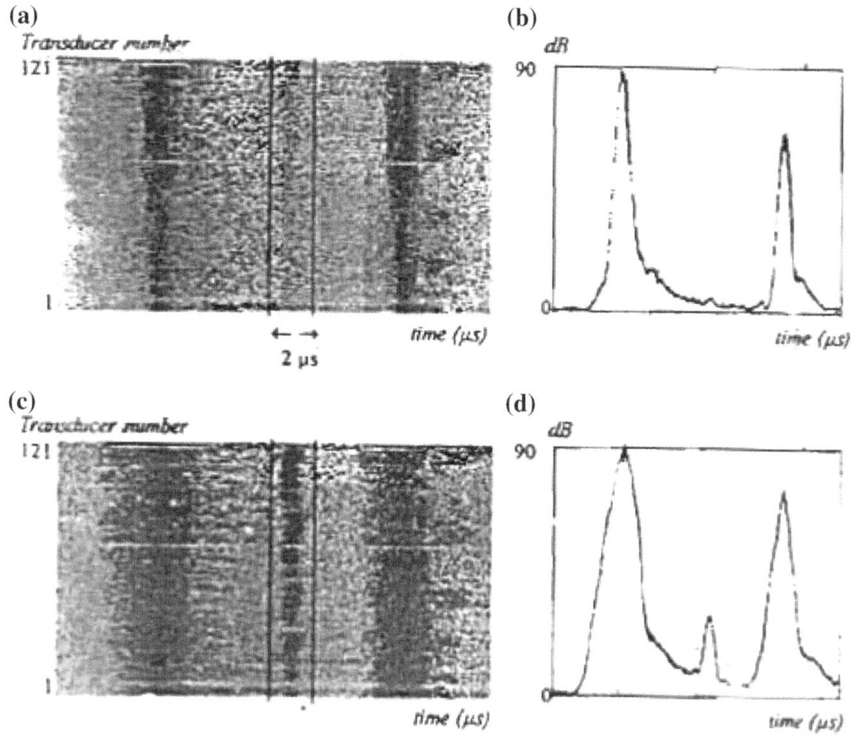

Fig. 12.4 Time reversal process on a zone containing pure speckle noise. **a** Grey-level B-Scan presentation of reception 1; **b** incoherent summation with logarithmic scale of reception 1; **c** grey-level B-Scan presentation of reception 2; **d** incoherent summation with logarithmic scale of reception 2 (from Chakroun et al. [18])

readily differentiated from the noise background. The fraction of the incident sound reflected depends on the magnitude and abruptness of the impedance contrast (and on the size and shape of the defect).

In the second step, a 2-μs (6 mm of titanium) time window is chosen after the front face echo, selecting a titanium section whose origin is located at a depth of 20 mm, same depth as the artificial hard-alpha defect.

In the third-step, the windowed data are time reversed and retransmitted. After propagations, the time-reversed wave focuses on the hard-alpha.

For the step, the echoes from the block are recorded and the corresponding data are shown in Fig. 12.3c. The echoes from the interfaces still exist but between them an oscillating line clearly appears. This line corresponds to the echoes from the hard-alpha received by the elements of the array. The defect signal can be readily differentiated from the noise background and the defect is detected. After a time reversal process, the signal-to-noise ratio increases. The amplitude and oscillation correspond to an off-axis defect whose wavefront intercepts the 2D array obliquely.

The technique is efficient whatever the position the defect in the incident beam. The TRM performs in real time; Fermat's surface matched to the relative positions between the TRM and the defect. The hard-alpha defect is automatically detected in a section of more than 1 cm^2 around the axis of the 2D TRM.

12.4.3 *Incoherent Summation*

A more compact presentation of the time reversal process can be implemented by adding the 121 logarithmic envelopes of the received signals (Fig. 12.3b, d). This sum generates a single-array output and the process is known as incoherent summation because the individual data are not in phase.

The incoherent summation for reception step numbers, Inc_s is determined according to the discrete summation

$$Inc_s[m] = \sum_{k=1}^{N} \log\left(R_{k,s}[m]\right) \quad 0 \le m \le M$$

The total output signal can be improved significantly by correcting the individual signals for the differences in arrival times. A summation of the shifted individual signals is performed to obtain a single combined signal for the array. Such a time-compensating process corresponds to a coherent summation and allows the echo level to be increased.

The coherent summation is much more efficient than the incoherent one.

12.4.4 *Time Record of Signals Coming from a Speckle Noise Zone*

In the second part of the experiment, the time reversal process is now evaluated with a time-reversed window located in a pure speckle noise zone. Figure 12.4c shows that the signal behaviour does not change after one time reversal process. They do not observe any wavefront appearing in the data. This is a different behaviour of time reversal on echographic speckle noise compared with hard-alpha signals. In titanium, the speckle noise comes from the microstructure whose dimension (a few μm) is small compared to the wavelength (1.2 mm for longitudinal waves). The time reversal process cannot refocus the energy on the acoustic source of the speckle noise. This is due to the loss of information on the small details of the titanium microstructure during propagation of the backscattered wave with a 5 MHz centre frequency.

12.4.5 The Iterative Technique

For a defect signal of low amplitude, multiple iterations can be implemented to improve the ability to focus on the defect, and thus increase the signal-to-noise ratio. The iterative time reversal mode can also be used to distinguish between speckle and defects.

To iterate the process, they start from reception 2. At this step, they define for each element k the new time reversal window $W_{k,2}[m]$. These windowed data are time reversed and stored in the corresponding transmit memory. The resulting emission, E_3, can be considered as the emitted wave of the next time reversal iteration. The operation can be repeated in an iterative loop, as shown schematically in Fig. 12.5.

Experimentally, two iterative loops are performed: a new time reversal operation is processed after reception 2, and new data from the medium are recorded during reception 3, R_3. In the following, they illustrate the results of the iterative time reversal operation for a zone containing a small defect and a zone containing only speckle noise.

A. Iterative Process for a Zone Containing a Hard-Alpha

In Fig. 12.6a, c, we observe that the signals recorded after one and two time-reversed operations, W_2 and W_3 are similar and show an identical waveform. Figure 12.6b, d shows the incoherent summation Inc_2 and Inc_3. Figure 12.7 represents the temporal signals received on transducer k, $W_{k,2}[m]$ (solid line) and $W_{k,3}[m]$ (dotted line). They are very similar up to an amplitude factor due to the amplification efficiency of the time reversal process. This result is linked to the fact that the echographic signals come from a coherent source of small dimension (the hard-alpha). We also note that the two waveforms are symmetric.

B. Iterative Process as a Pure Speckle Noise Zone

If the echoes come from a pure speckle noise zone (Fig. 12.8), a careful analysis of the signals W_2 and W_3 shows that they are random and completely uncorrelated. Figure 12.9 shows two waveforms receiving at receptions 2 (solid line) and 3 (dotted line) on the same transducer that was used to detect the signal shown in Fig. 12.7. They note that the two signals show quick variations and seem to be random signals. The two waveforms are uncorrelated and there is no amplification in this process. The time reversal has lost the information required to refocus on the small details of the microstructure, so there is no focusing effect.

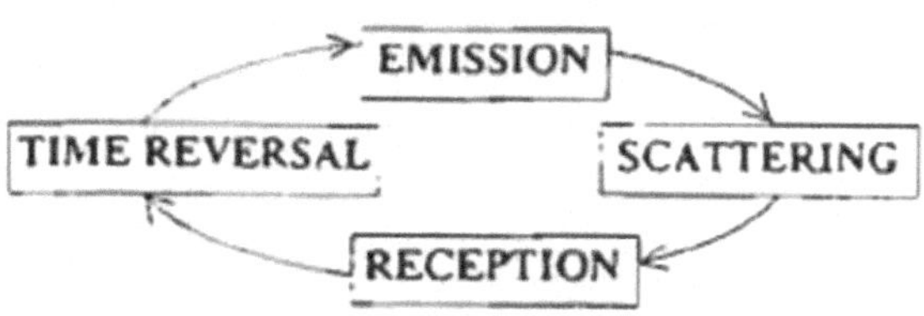

Fig. 12.5 Iterative mode loop (from Chakroun et al. [18])

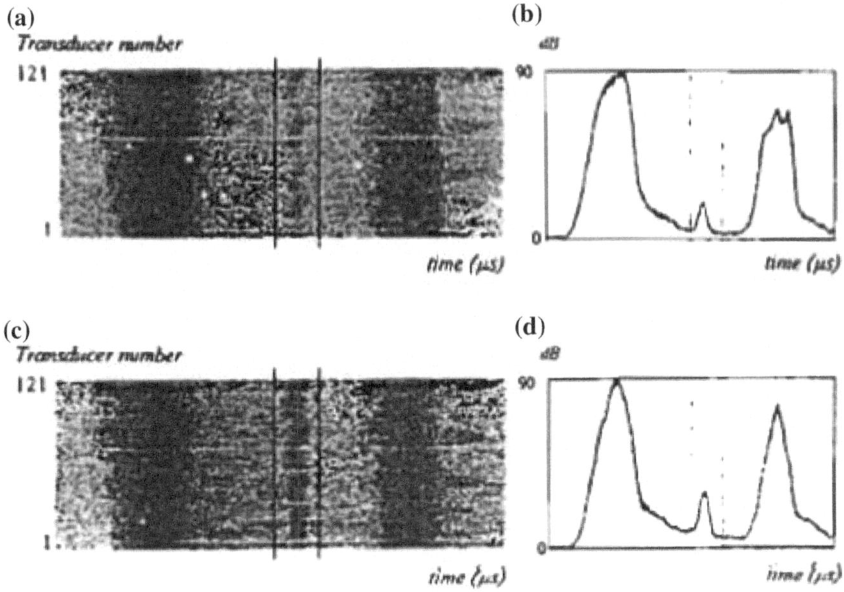

Fig. 12.6 Iterative time reversal process on a zone containing a defect: **a** grey-level B-Scan presentation of reception 2; **b** incoherent summation of reception 2; **c** grey-level B-Scan presentation of reception 3; **d** incoherent summation of reception 3 (from Chakroun et al. [18])

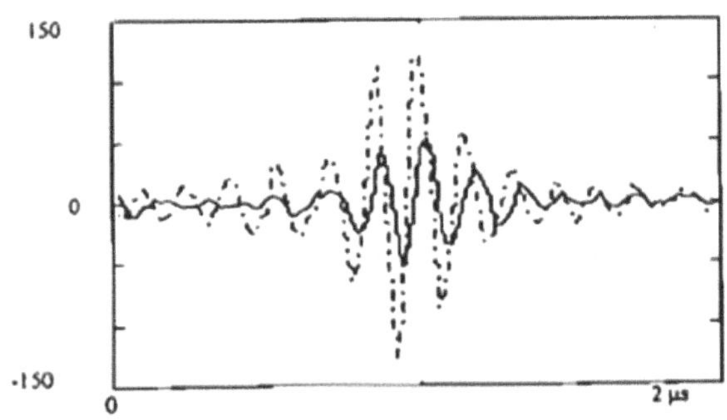

Fig. 12.7 Echoes of the hard-alpha after one (solid line) and two (dotted line) time reversal process (from Chakroun et al. [18])

So they concluded that the iterative time reversal process is an interesting tool for the inspection of noisy background media like titanium blocks for two reasons. First, this technique behaves differently depending on whether the inspection zone is coherent or not. This allows it to distinguish a defect from the surrounding material. Second, it improves the detection of small defects by increasing the signal-to-noise ratio.

When the iterative time reversal process is performed on a speckle noise zone, the resulting waveforms W_2 and W_3 are different. To understand these results, they

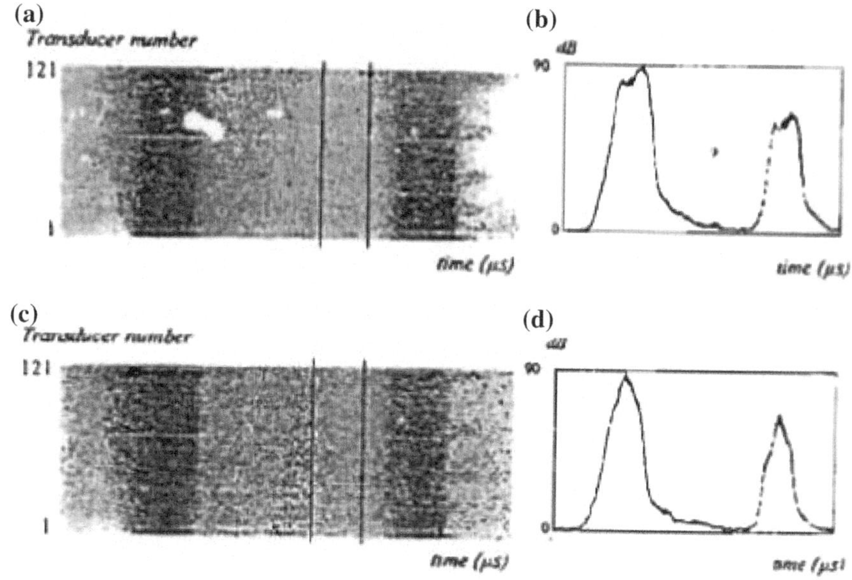

Fig. 12.8 Iterative time reversal process on a pure speckle noise zone: **a** grey-level B-Scan presentation of reception 2; **b** incoherent summation of reception 2; **c** grey-level B-Scan presentation of reception 3; **d** incoherent summation of reception 3 (from Chakroun et al. [18])

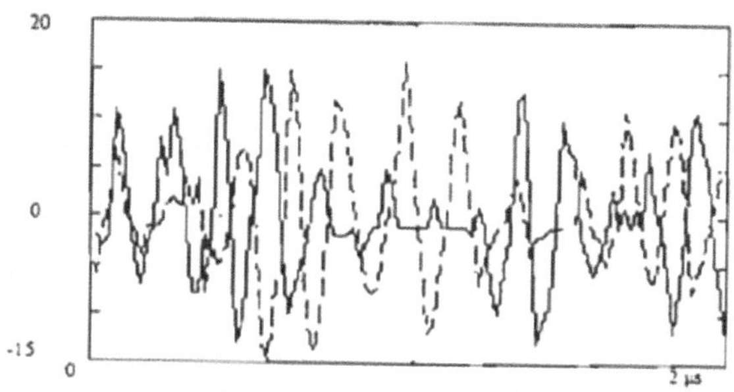

Fig. 12.9 Echoes of the speckle noise after one (solid line) and two (dotted line) time reversal process (from Chakroun et al. [18])

recall how the microstructure wave is generated. In a metal grain, the speed of sound depends upon the direction of propagation relative to the crystal lattice. Because the crystal lattices of two adjacent grains are usually not aligned with one another, there is a velocity and impedance contrast at a grain boundary. The reflected waves from the various grain boundaries encountered by the beam backpropagation in the direction of the 2D array, resulting in the microstructural noise. The microstructural noise depends upon the number of grains, as well as the relative position of the grains with respect to the acoustic beam.

Thus, TRM represents a real improvement in inspection of samples with high-ultrasound speckle noise levels. Experiments show the effectiveness of the time-reversed technique in locating a defect within a component that has a plane or a complex interface. With this technique, the signal-to-noise ratio is improved for two reasons: First, the time reversal process is a self-adaptive focusing technique that creates a focused wave matched to the defect shape to the propagation medium and to the geometry of the mirror and the medium. Second, the time reversal process reduces the amplitude of the speckle noise. Due to the loss of information during the propagation, the process cannot focus on microstructure whose dimension is small compared to the wavelength. It is also capable to detect a small defect located beneath an interface with complex geometry. The time reversal process of C-Scan imaging can be performed in real time with the signal processing process (incoherent summation and coherent summation) with DSP unit whose process takes only about 40 ms.

12.5 Application of Time Reversal Acoustics to Landmine or Buried Object Detection

12.5.1 Introduction

For the application of time reversal acoustic (TRA) to landmine or buried object detections, we first consider the feature of a landmine detection system which has a noncontact sensor which is used to measure ground motion, making it possible to sense motion directly above a landmine [20] (Fig. 12.10). While multiple wave types are generated by the system's excitation signal, the wave which is useful for the detection of landmine is the Rayleigh surface wave. Rayleigh wave propagates near the surface along the boundary between air and the soil and interacts with objects buried in the medium in the form of scattering of the Rayleigh wavefront off of the object. When the buried object is a landmine, due to its structure and the depth at which it is usually buried, the Rayleigh wave may excite a resonance in the layer of the soil between the surface and the flexible top of a landmine. This resonance enhances the surface displacements and is the primary detection for buried landmine [20].

Scattering off clutter objects in the medium causes the Rayleigh wave to become disorganized. If a large number of objects are present, the multiple scattering can interfere with the Rayleigh wave to the extent that it no longer effectively illuminates the buried landmine. Any resonance that is excited will be difficult to detect in the presence of the multiply scattered waves off objects in the medium. Time-reversed focusing method is useful here to focus energy to a specific location within the medium, irrespective of the presence of clutter. This allows one to focus energy to a certain spot in order to excite a resonance in any target that may be present there.

In this detection problem, sources are in the near field, only a few wavelengths from the targets and scattering objects. The seismic system differs significantly from

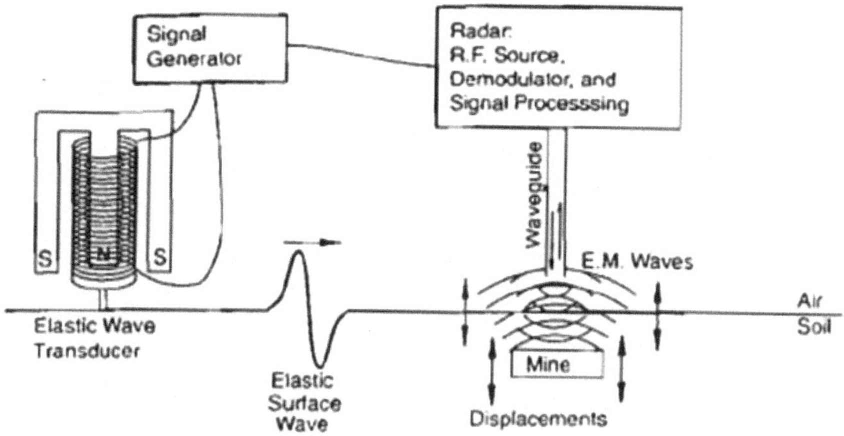

Fig. 12.10 Schematic of the elastic wave landmine detection system (from Norville and Scott 2005)

ultrasound system in that energy is coupled directly into the soil, rather than through a liquid conplant. The coupling of the transducer motion into the soil significantly alters the frequency response of the excited wave.

12.5.2 Theory

The acoustic wave equation for elastic waves propagation in solids is used:

$$\rho_s = \frac{\partial^2 \vec{u}}{\partial t^2} = (\lambda + 2\mu)(\nabla(\nabla \cdot \vec{u})) - \mu(\nabla \times (\nabla \times \vec{u})) \tag{12.23}$$

where

$\vec{u}$ displacement,
λ, μ Lamé constants of the medium and
ρ_s density of medium

Here, the external forces or body forces such as the gravitational force is neglected. It also assumes that the medium is lossless with respect to wave propagation.

Equation (13.23) consists only of second-order time derivatives. This means that if $\vec{u}(\vec{r}, t)$ is a solution to this equation, then $\vec{u}(\vec{r}, -t)$ must also be a solution to this equation. Because experimentally it is necessary to work with reversed time in a causal fashion, a finite time duration must be selected over which the equation will be considered. The formulation $\vec{u}(\vec{r}, T - t)$ over the interval $(T, 0)$ satisfies the causality requirement. If all the energy in the spatial region of interest is small outside of this time interval, then this solution should be almost exactly equal to $\vec{u}(\vec{r}, T - t)$.

A time reversal cavity is a three-dimensional (3D) surface which is constructed around a location of interest, usually a source location. All waves impinging on this surface are recorded, time reversed and retransmitted. Classical time-reversed focusing further simplifies this to a time reversal mirror (TRM) where only a portion of the time reversal cavity is realized. In the landmine detection application, the wave mode of interest is the Rayleigh surface wave which decays exponentially with depth. Though some energy is lost from mode conversion and from scattering objects in the soil, most of the Rayleigh wave's energy remains near the surface. Since landmines are buried near the surface and the energy in the Rayleigh wave is concentrated in that region, the landmine detection problem is a quasi-two-dimensional problem.

For the TRM, receivers are realized as a simple array. The array subtends some angle of the 3D surface that would be necessary to surround the focus point. The number and spacing of the array elements will have effects as grating lobes. The spot size of the focus point is also limited by that TRM aperture and diffraction effects proportional to wavelength [1].

12.5.3 Experimental Procedure

For the experimental implementation of time reversal focusing, elastic wave sources are located in an array $S_n = (x_{Sn}, y_{Sn} | n = 1, 2, \ldots, N)$ (Fig. 12.11). First, consider the effect of time reversal from a single source, S_n.

Step 1: Transmit an excitation signal, $\epsilon(t)$, from source S_n.
Step 2: Receive a signal, $f_n(t)$, at the desired focusing location R. Propagation through the medium is described by one Green's function, $G(S_n, R, t)$ such that

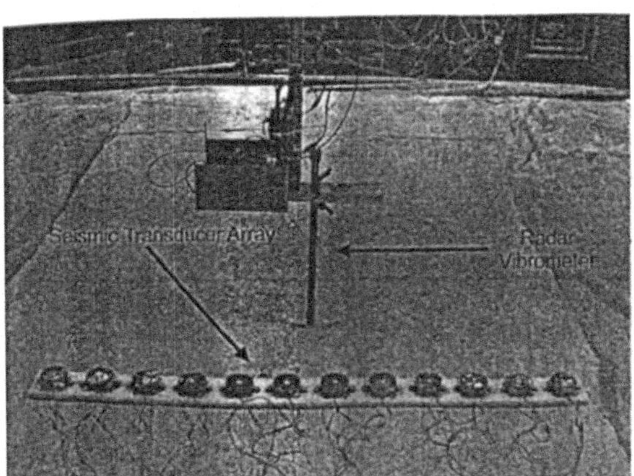

Fig. 12.11 Experimental facility. The seismic transducer array and the antenna are positioned over the sand bank (from Norville and Scott 2005) (Colour online)

$$f_n(t) = \epsilon(t) * G(S_n, R, t) \tag{12.24}$$

Step 3: Time reversed the received signal: $f(t) = f(-t)$.
Step 4: Transmit the time-reversed signal, $f(-t)$, from S_n and record at any location on the surface r, such that the signal recorded at r is

$$u_n(\vec{r}, t) = [\epsilon(-t) * G(S_n, R, -t)] * G(S_n, \vec{r}, t) \tag{12.25}$$

Using the associative property of convolution $u_n(\vec{r}, t)$ in Eq. (13.25) is the cross correlation of two Green's functions convoluted with the time reversal excitation functions.

When $r = R$, this becomes the autocorrelation function. This provides the mathematical explanation for the observed focusing of the signal at R.

Extending this to include additional transmitters in the array, one has

$$u(\vec{r}, t) = \sum_{n=1}^{N} [\epsilon(-t) * G(S_n, R, -t)] * G(S_n, \vec{r}, t) \tag{12.26}$$

In the experimental implementation of the time reversal method, steps 1–3 are preformed once for each transmitter S_n in the array. Step 4 is performed simultaneously for all transmitters $S_{1\ldots N}$.

The time reversal focusing method used in the experiment for landmine detection is different from the traditional time reversal focusing using a time reversal mirror (TRM), which requires either a source to be located at the desired focus location (R) or that an excitation be launched from the transducer array. In the case of landmine detection, it would be unwise to place a seismic source at a location where a landmine is believed to be buried.

For the traditional TRM, after the excitation is launched from the transducer array, reflection off a target at the focal location acts as a passive source. The reflections are recorded at the TRM, time reversed and retransmitted in the landmine or buried target detection problem, the signal reflected off a target is not strong enough to be significantly above the noise level. This makes it impractical to use reflected signal as a source for time reversal focusing.

The TRM for landmine detection relies on reciprocity of the propagation from the source to the focus point $G_n(R, S_n, t) = G_n(S_n, R, t)$. Applying reciprocity to U($\vec{r}$, t) will yield the autocorrelation function for the case of $r = R$. In the case of an anisotropic propagation medium, reciprocity may not be valid, and traditional TRM in implementation could fail to yield the autocorrelation function for the special case of $\hat{r} = R$.

12.5.4 Experimental Set-Up

An example of the experimental demonstration of the use of time reversal method for landmine detection is the system in the School of Electrical and Computer Engineering laboratory of the Georgia Institute of Technology (Fig. 12.11). Here, a large concrete wedge- shaped tank is filled with approximately 50 tons of damp compacted sand. The seismic waves are generated by an array of 12 electrodynamic shakers. A short metal bar foot is attached to each electrodynamic shaker. The shaker and metal foot are placed in contact with the sand and the 12.5 cm × 1.27 cm × 2.54 cm aluminium bar foot couple energy into the sand. Once the shakers are used to excite elastic waves in the sand bank, a noncontact electromagnetic sensor (radar vibrometer) is used to record the displacement of the surface of the ground. The vibrometer is scanned across the surface of the sound using a computer-controlled positioning system. The surface is sampled at 2-cm increments ($\Delta x = \Delta y = 2\,\text{cm}$) over a 1.2 m × 0.8 m area. The radar has a spot size of approximately 2 cm × 2 cm and records data at each location for 4,096 s at a sampling rate of 8 kHz. By making many measurements, each at a different location on the surface, the displacement of the entire scan region can be constructed synthetically. After the entire scan has been completed, a data array of displacement information is available, $D(x_i, y_j, t_k)$, where

$$\begin{aligned} x_i &= i\Delta x, i = 0, 1, \ldots, \frac{X\text{ cm}}{\Delta x} \\ y_i &= j\Delta y, j = 0, 1, \ldots, \frac{Y\text{ cm}}{\Delta y} \\ t_k &= k\Delta t, k = 0, 1, \ldots, \frac{T}{\Delta t} \end{aligned} \tag{12.27}$$

and where X and Y are the dimensions of the scan region and T is the duration of time for which each measurement is recorded.

A total of 113 rocks are buried in the sand bank in order to introduce inhomogeneities into the sand. The rocks are randomly distributed throughout the tank in location on both the surface and the burial depth.

12.5.5 Wiener Filter

In order to effectively illuminate the buried target using time reversal focusing, the excitation wave that reaches the target should be both broadband and compact in time. In addition, to being useful for time reversal focusing, a compact pulse allows for better separation of incident pulses and those reflected off a target. This separation is important for affiliated detection; it is important to ensure that the pulse that arrives at the target is broadband and temporarily compact. The practical way to do this is to design an inverse filter to restore the original response of the excitation signal. The propagating wave on the sand contains several different wave types, but the one of principal interest in the detection of buried targets is the Rayleigh surface wave. In order to most effectively design a filter that makes the Rayleigh wave temporarily compact and broadband, a signal processor [21] is used to extract the Rayleigh wave mode from the total propagating wave.

The Wiener filter is used here and designed so as to follow the conditions of the observed Rayleigh wave mode excitation signal resulting in a filtered excitation signal that is very similar to the desired temporarily compact, broadband excitation pulse.

The filter coefficients are determined by recording signal outputs in an uncluttered medium and extract the Rayleigh wave mode. This information is used to design the Wiener filter using the Stieglitz-McBride method [22]. The Stieglitz-McBride method iteratively minimizes the difference between the desired and designed filter impulse responses for computation of the optimal least mean-square wave coefficients.

12.5.6 Experimental Results

The experimental results are shown in Fig. 12.12. The presentation of the data displays the maximum displacement at each location over the entire time record. The image is formed by creating and displaying the array, $M(x, y)$, where

$$M(x_i, y_i) = \mathrm{Max} K|D(x_i, y_i, t_k)| \tag{12.28}$$

The result is presented as pseudo-colour graph of the magnitude of the vertical component of the particle displacement at the surface. The pseudo-colour scale used in the new graph is a 40 dB logarithmic scale from white (0 dB) to black (−40 dB).

The scattering effects of rocks and other objects are visible in the uniform excitation case (Fig. 12.12a). There are also areas of the scan region that are not effectively excited by the pulses, which will be referred to as shadow regions. An examination of the time-delayed excitation graph (Fig. 12.12b) shows that it focuses energy to a small area near the desired excitation point, but not on top of it. This is due to propagation velocity gradients in the medium and the presence of scattering objects. In a highly cluttered and inhomogeneous environment, time-delayed focusing fails to excite the focus point effectively. This makes time-delayed focusing excitation only marginally useful for detection of near surface targets in the presence of large-scale clutter and inhomogeneities.

The time reversal focusing result (Fig. 12.12c) is qualitatively similar to the time-delayed excitation focusing graph. A notable difference is that the maximum displacement occurs at the desired focus point in the time reversal case. The reason for this improvement is that the time reversal method inherently incorporates the

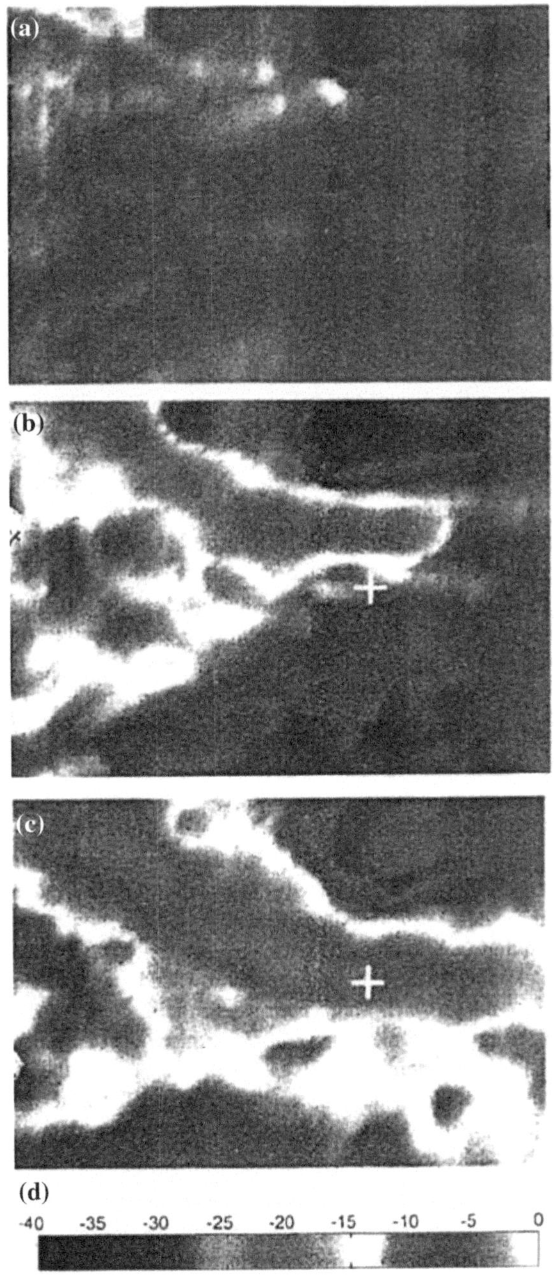

Fig. 12.12 (Colour online) Maximum displacement for Focus Point 1. Images are on a 40 dB pseudo-colour scale: 0 dB (white) to −40 dB (black). The desired focusing location is indicated by a white cross **a** uniform excitation, **b** time-delayed excitation, **c** time-reversed excitation, **d** colour-amplitude scale (from P.D. Norville and W.R. Scott [2005])

effects of scattering and variation in propagation velocity when calculating the time-reversed excitation pulse. It should also be noted that the displacement at the focus point is much larger than the displacement throughout the rest of the medium. This means that the interaction of the excitation pulse with the scattering objects has been significantly reduced in comparison to the uniform excitation case.

This is clear that time reversal focusing yields significantly advantages over the other excitation methods in the presence of clutter and variation in wavespeeds. This demonstration that time reversal can be one effective method of excitation in regions that are poorly illuminated by traditional excitation methods.

The condition under which time reversal shows the most dramatic improvements over other focusing methods is when a strong wavespeed gradient is present in the medium, normal to the direction of propagation. The specific advantage of time reversal over other methods is that time reversal requires no a priori knowledge of the characteristics of the background medium.

12.6 Application of Time Reversal Acoustics to Underwater Acoustics

An example of the application of time reversal acoustics to underwater acoustics has been carried out by Walker et al. [23] of the Marine Physical Laboratory of the Scripps Institution of Oceanography, University of California. A time reversal mirror (TRM) refocuses back at the original probe source position. A goal has been to refocus at different positions without model-based calculation. They presented a technique to refocus at different depths than the original probe source in a shallow ocean range-independent waveguide. In shallow ocean waveguide application, TR is often implemented using a vertical line array (VLA) of acoustic transducers covering some or all of the water column. The VLA is often referred to as a time reversal mirror (TRM) [24]. The requirement is to collect data from various ranges at a single depth, as from a moving broadband radiator, over a distance sufficient to construct the relevant frequency–wave number $(f - k)$ structure of the waveguide. With this information, it is then possible to focus at arbitrary depth at any of the ranges that the probe source data were taken. S.C. Walker et al. introduced a method for shifting the TR focus in depth from the initial probe source depth in the finite bandwidth model propagation regime.

12.7 Application of Time Reversal Acoustics to Nonlinear Acoustic Imaging

The advantage of nonlinear acoustic imaging is that it is more sensitive than linear acoustic imaging. It will be able to detect more minute changes in the material parameters such as to detect the defects in nondestructive material evaluation and the cancerous tissues in medical diagnosis. This can be illustrated by the considering the following mathematical relation:

$$y = ax + bx^2 + cx^3 \tag{12.29}$$

This shows that a 0.1 or 10% change in x will correspond to a 0.01 or 1% change in x^2 or 1% change is detectable. One of the methods of nonlinear acoustic imaging is the imaging of the biological structures in terms of nonlinear acoustic parameters of tissues that can provide tissue characterization sensitive to pathological changes. One of the methodologies is the B/A nonlinear parameter imaging []. This form of nonlinear acoustic imaging will require the interaction of focused ultrasound waves of two different frequencies which gives rise to the complex phase structure of the acoustics fields. This difficulties can be solved by the time reversal acoustics (TRA) due to the ability of the TRA to focus acoustic energy in a small volume with diameter close to that of half a wavelength. Also the advantage of the TRA lies in its capability to focus an ultrasound beam propagating in a heterogeneous medium. In addition, the TRA focusing will produce high acoustic intensity to enable the observation of various nonlinear acoustic effects useful for nondestructive evaluation (NDE) such as the second harmonic generation, modulation of sound by low-frequency vibration, amplitude-dependent resonance frequency shifts and amplitude-dependent internal friction.

Nonlinear TRA imaging will be scanning on a spot composed of two concentric focused acoustic fields of different frequencies generated by two synchronized TRA focusing systems. This acoustic intensity will provide information on the nonlinear parameters of the medium. The focal area will become a monopole source of the difference frequency waves.

12.8 Theory of Nonlinear Time Reversal Acoustics Nonlinear Acoustic Imaging

Here, the B/A nonlinear parameter of the medium will be derived from the measured signals of combination frequencies generated in the acoustic waves interaction zone. The Westervelt [25] equation of nonlinear acoustics will be used:

$$\nabla \mathrm{p} - \frac{1}{c^2}\frac{\partial^2}{\partial t^2}\, p = -\frac{\varepsilon}{\rho_0 c_0^4}\frac{\partial^2}{\partial t^2} p^2 \tag{12.30}$$

where

q nonlinear parameter $= 1 + \frac{B}{2A}$
p acoustic pressure
ρ_0 density of medium and
c_0 sound velocity in medium.

The nonlinearly generated secondary acoustic pressure p_s is obtained by perturbation method for a given primary field p_p:

$$\nabla p_s - \frac{1}{c^2}\frac{\partial^2}{\partial t^2}p_s = -\frac{\varepsilon}{\rho_0 c_0^4}\frac{\partial^2}{\partial t^2}p_p^2 \tag{12.31}$$

where $\mathrm{d}p_s = \frac{\rho_0}{4\pi r}\mathrm{q}(t - r/c_0)$ with $q = -\varepsilon/\rho_0^2 c_0^4 \frac{\partial^2}{\partial t^2}(p_p^2)$

For the interaction of two frequency waves (ω_1 and ω_2), the source density for the difference frequency q_- and the sum frequency q_+ is:

$$q_- = \frac{q(\omega_1 - \omega_2)^2}{2\rho_0^2 c_0^4} p_1 p_2*, \qquad q_+ = \frac{\varepsilon(\omega_1 + \omega_2)^2}{2\rho_0^2 c_0^4} p_1 p_2$$

The total radiated field, given by integral over the volume, can be written as:

$$p_\pm = \frac{q(\omega_1 \pm \omega_{2)}^2}{2\rho_0^2 c_0^4} \iiint_V p_{1,2}\frac{\exp(-ik_{1,2}r)}{r}\,\mathrm{d}V \tag{12.32}$$

with

$$k_{1,2} = (\omega_1 + \omega_2)/c_0 \tag{12.33}$$

The TRA-focused acoustic field is close to that of the spherically converged wave and can be written as:

$$p_{1,2} = \frac{\sin(k_{1,2}r)}{k_{1,2}r} \tag{12.34}$$

Substitution of (13.33) into (13.32) will produce:

$$p_\pm = \alpha(\omega_1 \pm \omega_2)^2 \varepsilon p_1(r = 0) p_2(\mathrm{r} = 0) d^3/24\rho_0 c_0^4 \mathrm{r}$$

where d = diameter of the focal area of interaction. The parameter α takes into account of the nonuniform distribution of the primary field. For difference frequency radiation where difference wavelength is much less than d, $\alpha = 1$.

12.9 The Experimental Set-Up to Analyse the Nonlinear Time Reversal Acoustic Waves

The experiment [26] will be conducted in a medium with spatially varying acoustical nonlinearity produced by ultrasound contrast agents. The nonlinear scattering from small focused area will provide the measurements of the nonlinear parameters in the region of the interaction. The TRA system will enable the scanning of the focus spot in the tested area to provide the nonlinear acoustic imaging.

Two synchronized TRA systems were used to focus ultrasonic waves with different frequencies 660 and 776 kHz onto a coinciding spot with spherical shape. Each TRA system as made of aluminium resonators consists of aluminium blocks with 5 piezotransducers glued onto the outside facet of the resonator. A cloud of microbubbles generated by electrolysis of water using a needle electrode was used to represent a medium with spatially varying acoustical nonlinearity. In the experiment, the acoustic signal is detected by a needle hydrophone, time reversed and sent back by the same TRA system in reradiation mode. Every step is controlled by the computer with an Ethernet connection.

The purpose of the experiment is to study the nonlinear interaction in the focal area of two-frequency TRA-focused ultrasound field. The information on the nonlinear parameter of the medium can be obtained from the measurements of the intensity of the sum frequency wave. The next step will be to scan the TRA focus point to provide nonlinear acoustic images of the scanned area. The experimental results show the nonlinear TRA signal distribution.

They show TRA system user interface in the stage of radiation of initial wave and TRA signal reradiation. They provide the TRA-focused signals and their spectra with and without the contrast agents. The nonlinear response is extracted with summation of responses when direct and inverted TRA signals are sent in the medium. Direct, inverted and direct + inverted TRA signals are shown. The nonlinear signal distortions are shown by the nonlinear response given by a black line. The comparison between nonlinearity generated at $f_1 = 660$ kHz alone, $f_2 = 771$ kHz alone and both at F_1 and F_2 shows that intermodulation component is generated at $f_1 + f_2 =$ 1436 kHz.

References

1. Fink, M.: Time reversal of ultrasonic fields-part I: basic principles. IEEE Trans. Ultrason. Ferroelect. Freq. Contr. **39**(5), 555–566 (1992)
2. Wu, F., Thomas, J.L., Fink, M.: Time-reversal of ultrasonic fields-part II: experimental results. IEEE Trans. Ultrason. Ferroelectr. Freq. Control **39**, 567–578 (1992)
3. Cassereau, D., Fink, M.: Time-reversal of ultrasonic fields-part III: theory of the closed time-reversal cavity. IEEE Trans. Ultrason. Ferroelectr. Control **39**, 579–592 (1992)
4. Cassereau, D., Fink, M.: Focusing with plane time-reversal mirrors: an efficient alternative to closed cavities. J. Acoust. Soc. Am. **94**, 2373–2386 (1993)
5. Draeger, C., Cassereau, D., Fink, M.: Theory of the time-reversal process in solids. J. Acoust. Soc. Am. **102**, 1289–1295 (1997)
6. Dowling, D.R., Jackson, D.R.: Phase conjugation in underwater acoustics. J. Acoust. Soc. Am. **89**, 171–181 (1990)
7. Hodgkiss, W.S., Song, H.C., Kuperman, W.A., Akal, T., Ferla, C., Jackson, D.R.: A long-range and variable focus phase-conjugation experiment in shallow water. J. Acoust. Soc. Am. **105**, 1597–1604 (1999)

8. Kuperman, W.A., Hodgkiss, W., Song, H.C., Akal, T., Ferla, C., and Jackson, D.R.: Phase conjugation in the ocean. J. Acoust. Soc. Am. **102**, 1–16 (1997)
9. Derode, A., Roux, P., Fink, M.: Robust acoustic time reversal with high-order multiple scattering. Phys. Rev. Lett. **75**(23), 4206–4209 (1995)
10. Fink, M.: Time reversed acoustics. Phys. Today **50**, 34–40 (1997)
11. Fink, M.: Time-reversed acoustics. Sci. Am. 91–97 (1999)
12. Blomgren, P., Papanicolaou, G., Zhao, H.: Super-resolution in time-reveral acoustics. J. Acoust. Soc. Am. **111**, 230–248 (2002)
13. Lehman, S.K., Devaney, A.J.: Transmission mode time-reversal super-resolution imaging. J. Acoust. Soc. Am. **113**, 2742–275 (2003)
14. Sutin, A.: Papers presented at the acoustical society of America's 75th anniversary meeting (2004), New York City, USA
15. von Ramm, O.T., Smith, S.W.: Beam steering with linear arrays. IEEE Trans. Biomed. Eng. **BME-30**, 438–452 (1983)
16. Turnbull, D.H., Foster,F.S. Beam steering with pulsed two-dimensional transducer arrays. IEEE Trans. Ultrason. Ferroelec. Freq. Contr. **38**, 320–333 (1991)
17. Smith, S.W., Pavy, H.G., von Ramm, O.T.: High-speed ultrasound volumetric imaging system-part 1: transducer design and beam steering. IEEE Trans. Ultrason. Ferroelec. Freq. Contr. **39**, 555–566 (1992)
18. Chakroun, N., Fink, M., Wu, F.: Time reversal processing in ultrasonic nondestructive testing. IEEE Trans. Ultrason. Ferroelec. Freq. Contr. **42**, 1087–1098 (1995)
19. Chakroun, N., Fink, M., Wu, F.: Ultrasonic non-destructive testing with time reversal mirrors. Proc. IEEE Ultrason. Symp. Tucson **2**, 809–814 (1992)
20. Scott Jr., W.R., Martin, J.S., Larson, G.D.: Experimental model for a seismic landmine detection system. IEEE Trans. Geosci. Remote Sens. **39**, 1155–1164 (2001)
21. Alam, M., McClellan, J.H., Scott, W.R., Jr.: Multi-channel spectrum analysis of surface waves. In: 37th Asilomar Conference on Signals, Systems and Computers (2003)
22. Steiglitz, K., McBride, L.: A technique for the identification of linear systems. IEEE Trans. Autom. Control **AC-10**, 461–464 (1965)
23. Walker, S.C., Roux, P., Kuperman, W.A.: Focal depth shifting of a time reversal mirror in a range-independent waveguide. J. Acoust. Soc. Am. **118**, 1341–1347 (2005)
24. KupermanW, A., Hodgkiss, W.S., Song, H.C.: Phase-conjugation in the ocean: experimental demonstration of an acoustic time-reversal mirror. J. Acoust. Soc. Am. **103**, 25–40 (1998)
25. Westervelt, P.J.: Parametric acoustic array. J. Acoust. Soc. Am. **35**(4), 535–537 (1963)
26. Dos Santos, S., Choi, B.K., Sutin, A., Sarvazyan, A.: Nonlinear imaging based on time reversal acoustic imaging. In: Proceedings of the 8ème Congrès Francais d'Acoustique, pp. 359–362 (2006)

Chapter 13
Negative Refraction, Acoustical Metamaterials and Acoustical Cloaking

Abstract An introduction is given. Then, there is an explanation of the limitation of Veselago's theory with the following coverage: introduction, gauge invariance of homogeneous electromagnetic wave equation, gauge invariance of acoustic field equation, acoustical cloaking, gauge invariance of nonlinear homogeneous acoustic wave equation; the discovery of negative refraction is a special case of coordinate transformations or a unified theory for negative refraction and cloaking, conclusions. This is followed by the multiple scattering approach to perfect acoustic lens. Acoustical cloaking is described with introduction, derivation of transformation acoustics and application to a specific example. Acoustic metamaterial and simultaneous negative mass density and negative bulk modulus is described. Then, there is acoustical cloaking based on nonlinear coordinate transformations. Acoustical cloaking of underwater objects and extension of double negativity to nonlinear acoustics are also provided.

13.1 Introduction

The phenomenon of negative refraction was first theoretically mentioned in Veselago's 1968 [1] paper with idea taken from Mandel'stam's [2] 1945 paper. This is the outcome of the two key parameters of electromagnetic waves: permittivity and permeability having negative values. The material having these properties is known as double-negative metamaterial (DNG). It is a special type of metamaterial. This paper did not receive much attention because it was not possible to fabricate double-negative metamaterial (DNG) although the other important form of metamaterial, the bandgap metamaterials such as photonic crystals and phononic crystals were fabricated much earlier in the 1980s and in the early 1990s, respectively, and photonic crystals were known more than 100 years ago by Lord Rayleigh. In 1999, John Pendry et al. [3] of Imperial College London successfully introduced the theoretical concept of split-ring resonator (SRR). This was a great contribution to the field of double-negative metamaterial (DNG) as his theoretical concept enabled Smith [4] of Duke University, USA, to successfully fabricate experimentally the double-negative

W. S. Gan, *Gauge Invariance Approach to Acoustic Fields*,
https://doi.org/10.1007/978-981-13-8751-7_13

metamaterial (DNG) using the concept of SRR. Then there was great increase in interests in double-negative metamaterials.

Metamaterials are artificial materials engineered to have properties that may not be found in nature. Metamaterials usually gain their properties from structure rather than composition, using small inhomogeneities to create effective macroscopic behaviour. It is a high-level form of composite material with periodic structure. Double-negative metamaterials (DNG) also known as left-handed material because the negative directions of the permittivity, the permeability and the Poynting vector will form an anti-clockwise rotation.

The phenomenon of negative refraction of left-handed symmetry can be considered and explained in the light of gauge theory. Maxwell's equations are the oldest gauge theory, and left-handed symmetry and the negative values of permeability and permittivity can be regarded as gauge condition. With the substitution of this gauge condition, there is no change in the form of the Maxwell's equations, or Maxwell's equations are invariant with respect to a set of negative values of permeability and permittivity. In a subsequent section, acoustical cloaking uses the concept of gauge invariance of the Maxwell's equations subjected to curvilinear coordinates transformation (used in general relativity) and metamaterial. This reconfirms that negative refraction and acoustical cloaking can be explained in terms of gauge invariance.

13.2 Limitation of Veselago's Theory

13.2.1 Introduction

Veselago [1] proposed in 1968 the concept of metamaterials for electromagnetic waves having simultaneous negative values of the permittivity and the permeability, or double negativity. In this section, we point out the limitation of Veselago's [1] theory. Veselago's [1] theory is based on the dispersion relation for isotropic solids. Our new approach or alternative approach is based on the gauge invariance approach to acoustic fields proposed by the author in 2007 [5]. We show that this approach can extend metamaterials from electromagnetic waves to acoustic waves from first principles without using analogy. Also it can remove the ambiguity of using the dispersion relation for the refractive index because both the positive and the negative signs for the refractive index simultaneously occur due to the square root sign, and this has to be justified. In addition, it is applicable to acoustical cloaking using coordinate transformations, a form of gauge invariance. We also discover parity invariance in acoustic field equations although the Maxwell's equations are known to be parity invariant. Electromagnetic metamaterials are materials with artificial electromagnetic properties defined by their subwavelength structure rather than their chemical composition. The left-handed metamaterials are a special type of metamaterial with parity p equals -1 and having the properties of negative permittivity and negative permeability and with the Poynting vector for energy flow in the opposite direction to that of wave propagation.

13.2.2 Gauge Invariance of Homogeneous Electromagnetic Wave Equation

Veselago [1] started with the dispersion relation for the propagation of electromagnetic wave in isotropic material. He considered the dispersion relation

$$k^2 = \frac{\omega^2}{c^2} n^2 \tag{13.1}$$

and

$$n^2 = \varepsilon \mu \tag{13.2}$$

where

k wave number,
ω frequency,
c wave velocity,
n index of refraction of the medium,
ε permittivity and
μ permeability.

Neglecting losses and regarding n, ε and μ as real numbers, it can be seen from (13.1) and (13.2) that a simultaneous change of the signs of ε and μ has no effect on these relations. That is (13.1) and (13.2) are also valid for $-\mu$ and $-\varepsilon$. He then shows that for $\varepsilon > 0$ and $\mu > 0$, then $\vec{E}$, $\vec{H}$ and k form a right-handed triplet of vectors and if $\varepsilon < 0$ and $\mu < 0$, they form a left-handed set where $\vec{E}$ = electric field and $\vec{H}$ = magnetic field. He then introduced direction cosines for the vectors $\vec{E}$, $\vec{H}$ and $\overrightarrow{k}$ and denote them by α_i, β_i and γ_i respectively, to characterize wave propagation in a medium:

$$G = \begin{pmatrix} \alpha_1 & \alpha_2 & \alpha_3 \\ \beta_1 & \beta_2 & \beta_3 \\ \gamma_1 & \gamma_2 & \gamma_3 \end{pmatrix} \tag{13.3}$$

The determinant of this matrix is equal to +1 if the vectors $\vec{E}$, $\vec{H}$ and $\overrightarrow{k}$ are a right-handed set and −1 if this set is left-handed. He then denoted this determinant by p and said that p characterized the "rightness" of the given medium. That is, the medium is "right-handed" if $p = +1$ and "left-handed" if $p = -1$.

In this section, I replace "rightness" by parity, and parity $= -1$ is for left-handed set and parity $= +1$ is for right-handed set. Parity is the language of gauge invariance. All physics laws obey parity invariance except the β decay in weak interaction

Also Poynting vector $\vec{S}$ which denotes energy flow and the parameter of highest interest in electromagnetic wave is given by

$$\vec{S} = \frac{c}{4\pi}\left[\hat{E} \wedge \vec{H}\right] \tag{13.4}$$

From (13.4), the vector $\vec{S}$ always forms a right-handed set with the vectors $\vec{E}$ and $\overrightarrow{H}$. Accordingly, for right-handed substances $\vec{S}$ and $\overrightarrow{k}$ are in the same direction and for left-handed substances they are in opposite directions [1]. Since the vector $\overrightarrow{k}$ is in the direction of the phase velocity, it is clear that left-handed substances with a so-called negative group velocity, which occurs in particular in anisotropic substances or when there is spatial dispersion. That is for left-handed substance or parity $= -1$, the Poynting vector is of opposite direction to the phase velocity or the direction of wave propagation. For right-handed substance or parity $= +1$, the Poynting vector is of same direction as the phase velocity.

In this section, we build left-handed metamaterial on the framework of gauge invariance as it covers the both characteristics of left-handed material: negative permeability and negative permittivity and the Poynting vector pointing in the opposite direction to the phase velocity direction.

Next we will show that the homogeneous electromagnetic wave equation is gauge invariant to negative permeability and negative permittivity.

For homogeneous medium, the electromagnetic wave equations can be given as:

$$\nabla^2\vec{E} - \frac{\varepsilon\mu}{c^2}\ddot{\vec{E}} = 0, \quad \nabla^2\vec{H} - \frac{\varepsilon\mu}{c^2}\ddot{\vec{H}} = 0 \tag{13.5}$$

We find that there is no change in the form of Eq. (13.5) if ε and μ are to be replaced by $-\varepsilon$ and $-\mu$. This shows the gauge invariance of Eq. (13.5) to negative values of permittivity and permeability.

13.2.3 Gauge Invariance of Acoustic Field Equations

The Helmholtz homogeneous acoustic wave equation is given by

$$\nabla^2 P + \frac{\omega^2}{\rho\kappa} p = 0 \tag{13.6}$$

where

p acoustic pressure,
ρ mass density and
κ bulk modulus.

Again we find that there is no change in the form of Eq. (13.6) if ρ and κ are replaced by $-\rho$ and $-\kappa$. This shows that the Helmholtz wave equation is gauge invariant to the negative values of ρ and κ.

Here we have extended the left-handed media to acoustics using gauge invariance formulation. We also discover the parity invariance of acoustic field equation instead of using the Veselago's theory. Left-handed media has parity equals -1.

13.2.4 Acoustical Cloaking

Acoustical cloaking deals with the deflection of bending of sound wave and the control of the propagation and direction of sound wave according to our specified direction.

Again Veselago's [1] theory of using dispersion relation is not relevant here. We use coordinate transformations, a form of gauge invariance. That is there is no change in the form of the acoustic field equation after the coordinate transformations, or the acoustic field equation is gauge invariant subjected to coordinate transformations.

As an illustration, we quoted the results from Cummer [6]:

Cummer [6] illustrated coordinate transformations for acoustics by using the linear acoustic equation for inviscid fluid:

$$j\omega p = -\kappa\nabla\cdot\vec{v},\quad j\omega p\vec{v} = -\nabla p \tag{13.7}$$

where

ω angular frequency,
v sound velocity.

Next he imposed a new set of curvilinear coordinates x', y' and z' on these equations. Using A as the Jacobian matrix of coordinates transformation from (x, y, z) to (x', y', z'), he expressed the gradient operation in the new primed coordinates as:

$$\nabla p = A^{\mathrm{T}}\nabla' p = A^{\mathrm{T}}\nabla' p' \tag{13.8}$$

and the divergence operation can be expressed as

$$\nabla\cdot\vec{v} = \det(A)\nabla'\cdot\frac{A}{\det(A)}\vec{v} = \det(A)\nabla'\cdot\vec{v'} \tag{13.9}$$

With these expressions, the original Eq. (13.7) can be written in the new coordinates as

$$j\omega p' = -\kappa\det(A)\nabla'\cdot\vec{v'}$$

$$j\omega\det(A)\left(A^{\mathrm{T}}\right)^{-1}\rho\left(A^{-1}\right)\vec{v'} = -\nabla' p' \tag{13.10}$$

which is in the same form as the original Eq. (13.7) but with the new medium parameters:

$$\kappa' = \det(A)\kappa, \quad \overline{\overline{p}} = \det(A)\left(A^{\mathrm{T}}\right)^{-1}\rho\left(A^{-1}\right) \tag{13.11}$$

Physically, this means that if one applies coordinate transformations to a solution to Eq. (13.7) and changes the medium properties according to Eq. (13.11), the transformed fields are a solution to the acoustic equations in the new medium.

13.2.5 Gauge Invariance of Nonlinear Homogeneous Acoustic Wave Equation

The nonlinear homogeneous acoustic wave equation up to the second order can be given as:

$$\kappa_1 \nabla^2 p + \kappa_2 \nabla^2 p\left(\frac{\partial p}{\partial x}\right) + \frac{\omega^2 p}{\rho} = 0$$

or

$$\rho\kappa_1 \nabla^2 p + \rho\kappa_2 \nabla^2 p\left(\frac{\partial p}{\partial x}\right) + \omega^2 p = 0 \tag{13.12}$$

where

κ_1 second-order bulk modulus and
κ_2 third-order bulk modulus.

Again if we replace ρ and κ_1, and κ_2 by $-\rho$, $-\kappa_1$, and $-\kappa_2$, there is no change in the form of Eq. (13.12). In another words, the nonlinear acoustic wave equation is gauge invariant to negative values of ρ, κ_1 and κ_2.

13.2.6 My Important Discovery of Negative Refraction Is a Special Case of Coordinate Transformations or a Unified Theory for Negative Refraction and Cloaking

Here we are considering both cloaking and negative refraction under the umbrella theory of coordinate transformations or gauge invariance of the form of equations under coordinate transformations. This is a pattern of nature and is applicable to all equations of physics covering both Maxwell's equations and the acoustic equation of motion. When the determinant of the direction cosine matrix (or transformation matrix) equals -1, one will have negative refraction or parity equals -1. Also when multiplying the original permittivity and the original permeability by the determinant value of -1 will produce negative values of the permittivity and the permeability. This shows that negative refraction is a special case of coordinate transformations used in cloaking problem when the determinant of the transformation matrix equals -1. This can be illustrated as follows:

$$\begin{pmatrix} v'_x \\ v'_y \\ v'_z \end{pmatrix} = \begin{pmatrix} \alpha_1 & \alpha_2 & \alpha_3 \\ \beta_1 & \beta_2 & \beta_3 \\ \gamma_1 & \gamma_2 & \gamma_3 \end{pmatrix} \begin{pmatrix} v_x \\ v_y \\ v_z \end{pmatrix} \tag{13.13}$$

When the determinant of the direction cosine matrix on the right-hand side of (13.13) equals -1, we have

$$\overrightarrow{v'} = -\vec{v} \tag{13.14}$$

Replacing the vectors by the examples of permeability and permittivity, we will have

$$\overrightarrow{\mu'_{lj}} = -\overrightarrow{\mu'_{lj}} \quad \text{and} \quad \overrightarrow{\varepsilon'_{il}} = -\overrightarrow{\varepsilon_{il}} \tag{13.15}$$

This shows that negative refraction also produces negative permeability and negative permittivity.

Since this gauge invariance of the form of equation is a pattern of nature of all physics equations, it is also applicable to the acoustic case where the equivalence of the permittivity and permeability are the mass density and the bulk modulus or compressibility.

This also shows that cloaking material or component will become the lens in the special case of negative refraction, and refraction is a special case of cloaking or the bending of light wave or sound wave when the path of wave propagation becomes linear from nonlinearity.

This shows that gauge invariance has a broader coverage and applications than Veselago's [1] dispersion relation.

Further in Chap. 14 under new acoustics, we also introduce reflection invariance (or right-left symmetry) to explain negative refraction. In fact, $-\mu$ and $-\varepsilon$ can be considered as the mirror image of μ and ε and $-\rho$ and $-\kappa$ can be considered as the mirror image of ρ and κ. Again here the concept of coordinate transformation is used.

Of course, it should be also mentioned here that gauge invariance approach to negative refraction removes the ambiguity caused by using the dispersion relation. There are both positive and negative signs which occur simultaneously due to the square root sign of the dispersion relation, and this has to be justified.

13.2.7 Conclusions

The above evidence shows that Veselago's [1] theory is applicable only to electromagnetic waves and for isotropic materials and for the special case of double negativity and for linear case. Gauge invariance approach on the other hand has broader applications even to acoustic waves, to anisotropic materials, to cloaking problems, to negative refraction and to nonlinear acoustics. It also has the important contribution of removing the ambiguity occurred of whether to use the positive or the negative sign of the dispersion relation.

In fact, after Smith's [4] and Pendry's [3] accomplishments with metamaterials, Veselago realized that the most important contribution of his original 1968 paper is not that a composite material can be designed to produce a negative refraction, but that a composite material can be designed to produce any value for permittivity and permeability. This reconfirms my important discovery that negative refraction or double negativity in permittivity and permeability is only a special case of the general case of cloaking using coordinate transformations where a composite material can be designed to produce any value for permittivity and permeability.

13.3 Multiple Scattering Approach to Perfect Acoustic Lens

The multiple scattering theory (MST) usually known as the Korringa–Kohn–Rostoker (KKR) approach [7, 8] was developed mainly for the calculation of electronic band structures although it is originated from the study of classical waves including acoustic waves used by Liu et al. [9] to calculate the propagation of sound waves in periodic structures such as phononic crystals. The phononic crystals in this case are stainless steel balls immersed in water. They found theoretical and experimental agreement using ultrasound experiment of the observation of a sizable directional stopband in the transmission along (001) centred at about 0.65 units, coincides with unexpectedly directional gap along the $\Gamma - x$ direction in the band structure. In the transmission along (111), they observed a narrow stopband at about 0.65 units, corresponding to the small gap at the L point in the band structure at the same frequency.

Other works on the studies on the existence and properties of phonon bandgap are [10–13]. These are due to Bragg's scattering when the sound wavelength is comparable with the lattice constants. This leads to frequency bands where wave propagation is forbidden. This enables the understanding of how to achieve large complete bandgaps in physically realizable materials and the mechanism of wave transport at band frequencies due to tunnelling [14]. Also there has been relatively less attention paid to investigate how periodicity affects wave propagations over a wide range of frequencies outside the bandgaps where novel refraction, diffraction and focusing effects may be possible.

At low sound frequencies, an effective continuum or medium approximation can be used to study the wave properties and accurately predict the wavespeed. In this frequency range, there is much common with the properties of low-frequency phonons in atomic crystals, where phonon-focusing phenomena have been systematically studied [15]. However, at higher frequencies, much less is known about the behaviour in pass bands where the wavelengths can be much less than the lattice constant. Suxia Yang et al. have addressed this problem by theoretically and experimentally investigating the character of wave pattern and propagation in a 3D phonon crystal at frequency above the first complete bandgap. They showed how a dramatic variations in wave propagation with both frequency and propagation direction can lead to novel focusing phenomena associated with large negative refraction. This is a different approach to negative refraction from that of Veselago's work for the electromagnetic wave based on negative values of permittivity and permeability. They demonstrated the effect of negative refraction experimentally by using ultrasound technique to image the transmitted wave field and show that a flat crystal can focus a diverging incident beam into a sharp focal spot that can be seen remarkably far from the crystal.

They also calculated the field pattern theoretically using a Fourier imaging technique in which wave propagation through the crystal is accurately described by the 3D equifrequency surfaces predicted from the multiple scattering theory (MST) [16]. Their theoretical results also give an excellent explanation of the experimental data, showing how wave physics in the regime can be accurately modelled and how the theoretical structures on the equifrequency surfaces of phonon crystal can give rise to potential applications.

Zhang and Liu [17] first discussed the issue of negative refraction for acoustic waves in phononic crystals. They also repeated the observation of the negative refraction of acoustic wave in phononic crystals, occurring at the frequencies with $\vec{S} \cdot \vec{k} > 0$ where $\vec{S}$ represents the Poynting vector. They considered a 2D phononic crystal consisting of infinite-length "rigid" or liquid cylinders embedded in a background which has been studied extensively in Ref. [19–21]. Two types of phononic crystals were used by them. One is steel cylinders in air background, and the other is water cylinders in mercury background. The band structures of these two types of phononic crystals were plotted in Fig. 13.1a, b, respectively. Both of them were calculated by the MST (or Korringa–Kohn–Rostoker method given in Ref. [23]).

To visualize and analyse refractive effects of the acoustic wave when it hits the above phononic crystal interfaces, Zhang et al. [17] investigated the equifrequency surfaces (EFS) of the band structures just like the case for the electromagnetic waves in the photonic crystals because the gradient vector of constant-frequency contours in k-space give the group velocities of the phononic modes. Hence, the propagation direction of energy velocity of acoustic wave can be reduced from them. The EFS can also be calculated using the MST or the Korringa–Kohn–Rostoker method. The features of the EFS for these two kinds of system within the first band are similar. Thus, only the results of water–mercury system with $R = 0.4\,a$ in given in Fig. 13.2. The equifrequency surface contours at several relevant frequencies such as 0.05, 0.1, 0.2, 0.235 and 0.27 are demonstrated. It is clear that the lowest band has $\vec{S} \cdot \vec{k} > 0$ everywhere within the first Brillouin zone, meaning that the group velocity is never opposite to the phase velocity. The 0.05 and 0.1 contours are very close to a perfect circle, and the group velocity at any point of the contour is collimated with the $\vec{k}$

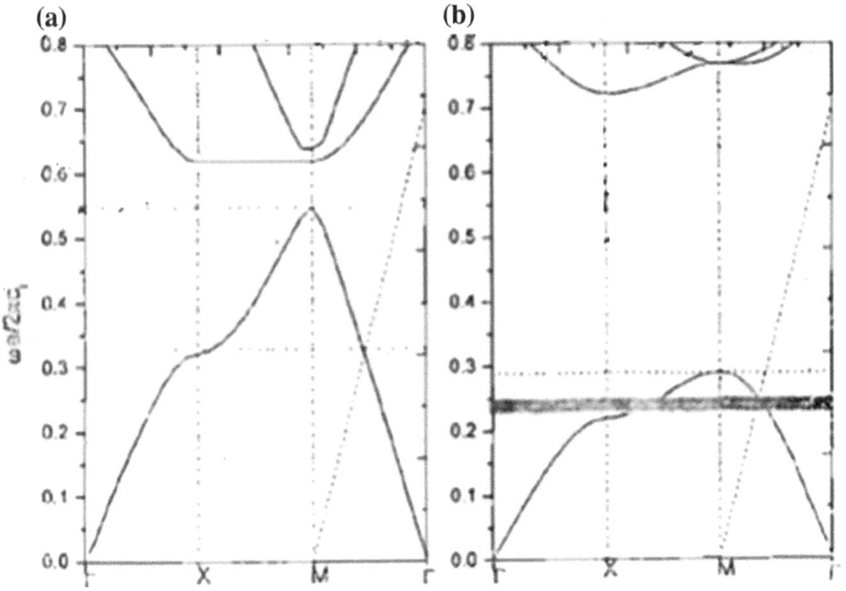

Fig. 13.1 **a** Acoustic band structures for a square lattice of steel cylinders in air background, with cylinder radius $R = 0.36a$. **b** The acoustic band structures for a square lattice of water cylinders in mercury background with cylinder radius $R = 0.4a$. The light line shifted to M is shown in dashed line. Dot-dashed lines mark the region for negative refraction, and the shadow represents the AANR region (From Zhang and Liu [17])

vector, indicating that the crystal behaves like an effective homogeneous medium at these two long wavelengths. The 0.2 contour is a little bit distorted from a circle, and the 0.235 contour is convex around the M point due to a negative phononic "effective band". The conservations of the component along the surface of refraction would result in the negative refractions' effect in some frequency region, marked as dotted lines in Fig. 13.1.

Furthermore, according to the analysis approach of [18, 19], the required condition for all-angle negative refraction (AANR) effect in some cases can be observed. Under these conditions, an acoustic beam incident on the ΓM surface with various incident angles will couple to a single Bloch's mode that propagates into this crystal on the negative side of the boundary normal. Therefore, we can define a frequency region for the AANR by using these criteria.

From Fig. 13.1a, we noted that the AANR region is absent in the steel–air system, although the negative refractive region is very large. However, in the water–mercury system the AANR region exists within the range of about 63 near $\omega = 0.24(2\pi c_l/a)$ (shadow region in Fig. 13.1b). This point differs for the two kinds of system. This difference is very important for the superlensing and focusing of acoustic waves in phononic crystals.

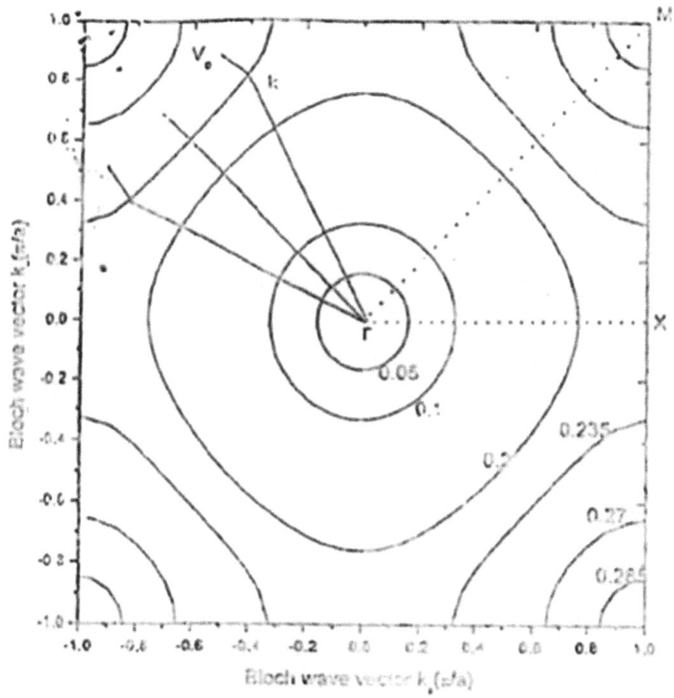

Fig. 13.2 Several constant-frequency contours of the first band of the 2D phononic crystal, which is composed of a square lattice of water cylinders in mercury background with $R = 0.4a$. The numbers in the figure mark the frequencies in unit of $2\pi c_l$/a (From Zhang and Liu [17])

In order to test this theoretical analysis, Zhang et al. [17] performed a numerical simulation to the two phononic crystals system based on the MST [20]. They used a 30°-wedged sample which consisted of 238 water cylinders of $R = 0.4\,a$ in the mercury background with a square array. The shape of the sample and an illustration of the refraction process are shown on the top of Fig. 13.3. The black frame marked the size of the sample. The wedged surface was the (11) surface when a slit beam of frequency $\omega = 0.235(2\pi c_l/a)$ with a half-width $wl = 2a$ incident normal to the left surface of the sample, it transports along the direction of incidence wave until it meets the wedge (11) interface of the sample, and then a part of it will refract outside of the sample and the other reflect inside. There are two possibilities for the refracted wave. It may travel on the right side (positive refraction) or left side (negative refraction) of the surface normal. The simulation results are plotted in Fig. 13.3. The field energy pattern of the incidence and refraction are shown in the figure. The arrows and text illustrate the various beam directions. It can be clearly seen that the density flux of the refractive wave outside of the sample travels on the negative refraction side of the surface normal. The refraction angle is consistent with the estimation from the wave-vector space in Fig. 13.2. The simulation results show clearly that the negative

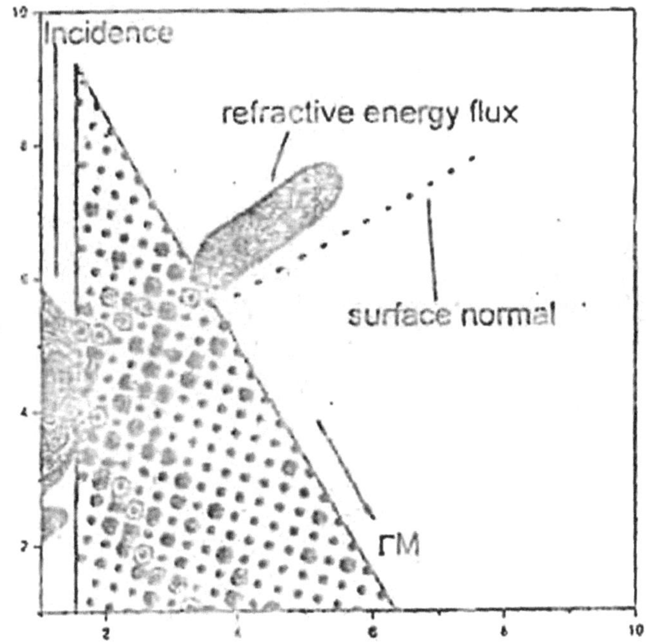

Fig. 13.3 Simulation of negative refraction. The boundaries of the sample are marked with black frame. The intensity of pressure field for incidence and refraction is shown in different shadows. A wedged sample considered here consists of water cylinders in mercury background with $R = 0.4a$ as shown on top of the figure. The frequency of incident wave is $\omega = 0.235(2\pi c_l/\text{a})$ (From Zhang and Liu [17])

refraction of the acoustic wave exists in the first band for the case with $\vec{S} \cdot \vec{k} > 0$. Similar phenomena have also been demonstrated in the steel–air system.

The concept of perfect lens or micro superlens has been designed using the concept of negative refraction [1, 21] and fabricated with 2D photonic crystals [19].

Such a superlens can focus a point source on one side of the lens into a real point image on the other side even for the case of a parallel sided slab of material. The advantage of the superlens or perfect lens is the capability to defeat the diffraction limit or Rayleigh resolution criterion of wavelength divided by two. Such an image can be realized by flat slab instead of curved shapes and thus fabrications can be easier in principle. Zhang and Liu [17] demonstrated the design of such a perfect lens for sound waves which possesses the same advantage as that of optical system. They used a slab of the sample with $40a$ width and six layers thick. A continuous wave point source is placed at a distance $1.0a$ from the left surface of the slab. The frequency of the incident wave emitting from such a point source is $\omega = 0.24(2\pi c_l/a)$, chosen to be within the region where all-angle negative refractions may occur (Fig. 13.1b).

The MST method is used to calculate the propagations of an acoustic wave in such a system. The typical results of field pattern of pressure wave and their images

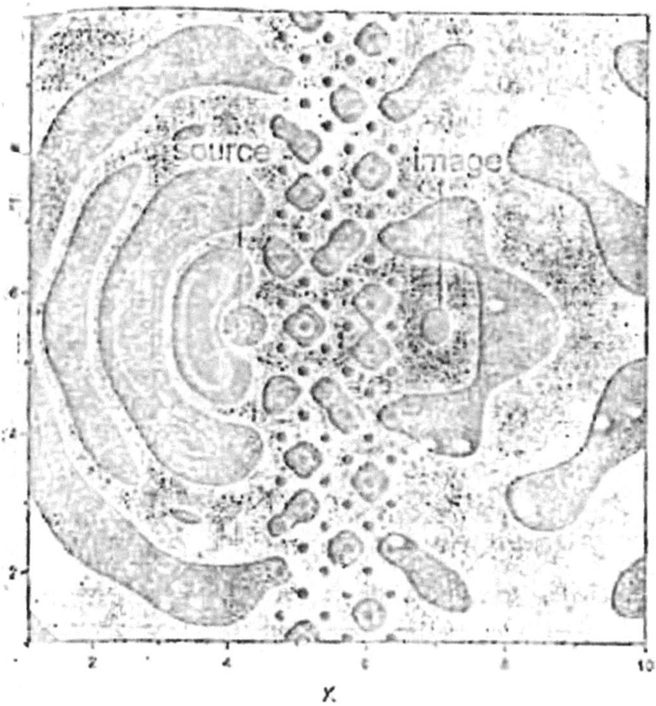

Fig. 13.4 Field pattern of pressure wave of a point source and its image across a six-layer slab at frequency $\omega = 0.24(2\pi c_l/a)$. The system considered here consists of some water cylinders in mercury background with $R = 0.4a$. Dark and bright regions correspond to negative and positive values, respectively () [From Zhang and Liu [17])

across the slab sample are given in Fig. 13.4. The geometry of the phononic crystal slab is also displayed. One can find quite a high-quality image formed in the opposite side of the slab. A closer look at the data reveals a transverse size (full size at half-maximum) of the image spot as $0.6a$ (or 0.14λ) at a distance of $1.0a$ from the right surface of the slab. The focusing size of the image depends on certain parameters such as the thickness of the slab and the distance between the source and the slab which is similar to the case for the optical system. The tuning of these parameters will produce a clearer acoustic image.

They also studied the effect on the image quality when the frequency of the sound wave is outside of the AANR region and system without the AANR region such as steel–air system. For these cases, the focusing phenomena are degraded. These show that the AANR is very important for the image formations.

This shows that negative refraction for acoustic wave in the 2D phonons crystal exists in a manner similar to that of optics.

13.4 Acoustical Cloaking

A. Introduction

Acoustical cloaking can be classified as a form of acoustical imaging because by placing a metamaterial acoustical cloak on the object to be cloaked it will render its disappearance from one sight. The concept of acoustical cloaking also extended from electromagnetic cloaking [22, 23]. Electromagnetic cloaking uses concepts of gauge invariance from general relativity, that is the form of the Maxwell's equations remain unchanged under arbitrary coordinate transformations with transformed permittivity and permeability values which are scaled by a common factor. Because of the nature of negative refraction of metamaterial, by cloaking the object with a metamaterial, the light rays will be deflected, stretched and bended and guided around the object and returned to their original trajectory.

However, due to the dispersion nature of the light, the cloaking effect is specific only to a single frequency and not broadband.

The concept of acoustical cloaking was extended to acoustics by Milton et al. in 2006 [24] and by Cummer and Schurig, in 2007 [6]. The analysis by Milton et al. [24] indicated that the coordinate transformations approach cannot be extended to elastodynamic waves in solids in the fully general case or even for the special case of compressional waves in a fluid. However, a scattering theory analysis has shown that the cloaking solution exists for acoustic waves in fluids as three dimensions (Cummer et al. [30]) and by analogies with electromagnetics. It has been shown that 2D acoustic waves [6] and 3D acoustic waves [25] can be made transformation invariant. The material parameters required to implement acoustic coordinate changes have also been obtained by Greenleaf et al. [26].

It has to be noted that the phenomenon of acoustical cloaking cannot be transplanted blindly from electromagnetic cloaking using analogy. As shown in section two of this chapter, Veselago's [1] theory is not applicable to acoustic waves and even for electromagnetic waves is valid only to isotropic case and not for anisotropic cloaking material which most cloaking materials are made of. Also the acoustic metamaterial has to be derived using the theory of elasticity and not from dispersion relation as what used to derive the Veselago's [1] negative permeability and negative permittivity. Our gauge invariance approach can provide better physical understanding of negative refraction and cloaking. We also noted that acoustic negative refraction can be obtained from multiple scattering theory (MST) besides the approach of negative mass density and negative bulk modulus. This also confirms negative refraction is a form of multiple scattering. The above analysis is also given in section two of this chapter.

Our idea of objection to use analogy between acoustic wave and electromagnetic wave is supported by Cummer et al. [27]. They pointed out that demonstrating the invariance through analogy of acoustic wave with electromagnetic wave masks some of the physics of the transformations approach particularly how vectors such as particle velocity and the pressure gradient change under transformations. Through an analysis of how power flow and constant phase surfaces must transform for completely general waves, they show that the velocity vector in acoustics must transform in a different way than the $\vec{E}$ and $\vec{H}$ vectors in electromagnetics. This explains why Milton's et al. [24] elastodynamics analysis which assumed that the acoustic velocity transforms like $\overrightarrow{E}$ and $\overrightarrow{H}$ did not result in acoustic equation transformation invariance. We feel that this further shows the intrinsic elastic properties of acoustic wave as different from the electromagnetics. The treatment of negative refraction using theory of elasticity approach by Lee et al. [28] and Gan's analysis on the gauge invariance of acoustic fields [5] further confirm this. An example of the fabricating of acoustical cloak is given by Cheng et al. [29].

B. Derivation of Transformation Acoustics

Here we follow approach of Cummer et al. [30]. The fluids' version of the linear acoustic field equations will be used:

$$\nabla p = i\omega\rho(\vec{r})\rho_0\vec{v} \tag{13.16}$$

$$i\omega p = \kappa(\vec{r})\kappa_0\nabla\cdot\vec{v} \tag{13.17}$$

where $\rho(\vec{r})$ and $\kappa(\vec{r})$ are the normalized density and bulk modulus respectively of the medium and are coordinate transform invariant. We will demonstrate how the acoustic $\vec{v}$ must transform by considering $\vec{v}$ in a nonorthogonal coordinate system described by coordinate q_1, q_2 and q_3 with unit vectors $\hat{u}_1$,$\hat{u}_2$ and $\hat{u}_3$ respectively. Following Pendry et al. [22] and letting $i = 1, 2, 3$

$$Q_i^2 = \left(\frac{\partial x}{\partial q_i}\right)^2 + \left(\frac{\partial y}{\partial q_i}\right)^2 + \left(\frac{\partial z}{\partial q_i}\right)^2 \tag{13.18}$$

Figure 13.5 shows what happens when we apply the divergence theorem to an infinitesimal volume in this nonorthogonal coordinate system.

Deriving the net outward flux of $\vec{v}$ from this volume and setting it equal to the divergence of $\vec{v}$ times the infinitesimal volume, it can be shown that

$$\begin{aligned}(\nabla\cdot\vec{v})Q_1Q_2Q_3\left|\hat{u}_1\cdot(\hat{u}_2\times\hat{u}_3)\right| &= \frac{\partial}{\partial q_1}[Q_2Q_3\vec{v}\cdot(\hat{u}_2\times\hat{u}_3)] \\ &+ \frac{\partial}{\partial q_2}[Q_1Q_3\vec{v}\cdot(\hat{u}_1\times\hat{u}_3)] + \frac{\partial}{\partial q_3}[Q_1Q_2\vec{v}\cdot(\hat{u}_1\times\hat{u}_2)]\end{aligned} \tag{13.19}$$

Let $V_{frac} = \left|\hat{u}_1\cdot\left(\hat{u}_2\times\hat{u}_3\right)\right|$ because this is the fraction by which a unit volume is compressed by the coordinate nonorthogonality and we use the conventional superscript (subscript) notation for contravariant (covariant) vector components using

$$\vec{v}\cdot\left(\hat{u}_2\times\hat{u}_3\right) = v^1\hat{u}_1\times\left(\hat{u}_2\times\hat{u}_3\right) \tag{13.20}$$

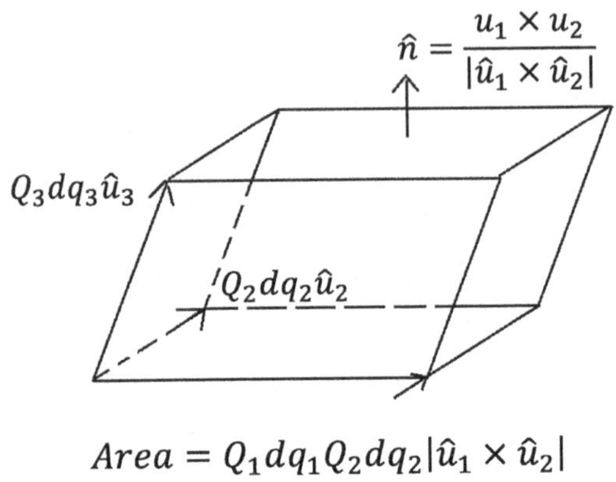

Fig. 13.5 Parallelepiped that defines an infinitesimal volume in the transformed coordinates. The area and unit normal of each face enters in the calculation of the net flux of a vector out of this volume (From Cummer et al. 30)

Equation (4) can be rewritten as

$$(\nabla \times \vec{v})Q_1Q_2Q_3V_{\text{frac}} = \frac{\partial}{\partial q_1}\left(Q_2Q_3V_{\text{frac}}v^1\right) + \frac{\partial}{\partial q_2}\left(Q_1Q_3V_{\text{frac}}v^2\right) + \frac{\partial}{\partial q_3}\left(Q_1Q_2V_{\text{frac}}v^3\right) \tag{13.21}$$

Noting that the divergence in the transformed coordinates is defined by $\nabla_q \cdot \vec{v} = \frac{\partial v^1}{\partial q_1} + \frac{\partial v^2}{\partial q_2} + \frac{\partial v^3}{\partial q_3}$, we can write

$$\begin{aligned}(\nabla \cdot \vec{v})Q_1Q_2Q_3V_{\text{frac}} &= \nabla_q \cdot (V_{\text{frac}}\overline{\overline{Q}}_{\text{per}}\left[v^1v^2v^3\right]^{\mathrm{T}} \\ &= \nabla_q \cdot \tilde{v}\end{aligned} \tag{13.22}$$

where

$$\overline{\overline{Q}}_{\text{per}} = \begin{bmatrix} Q_2Q_3 & 0 & 0 \\ 0 & Q_1Q_3 & 0 \\ 0 & 0 & Q_1Q_2 \end{bmatrix} \tag{13.23}$$

and the transformed velocity vector $\tilde{v}$ is given by

$$\tilde{v} = V_{\text{frac}}\overline{\overline{Q}}_{\text{per}}\left[v^1v^2v^3\right]^{\mathrm{T}} \tag{13.24}$$

The per subscript on the tensor $\overline{\overline{Q}}_{\text{per}}$ is to denote that the diagonal elements transform each vector component by the product of the coordinate scaling factors perpendicular (more general, not parallel, for the case of nonorthogonal coordinates) to

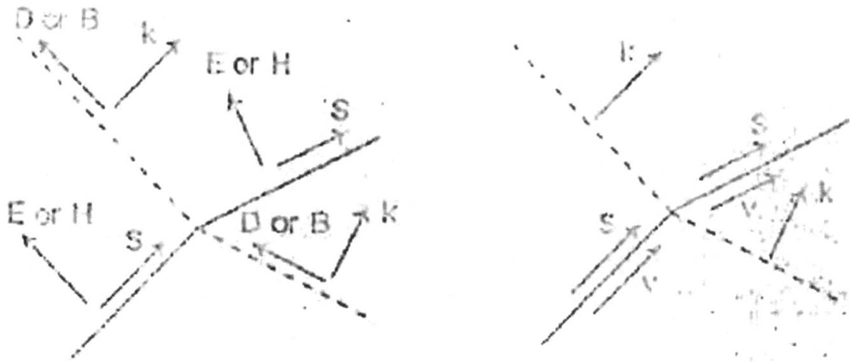

Fig. 13.6 Transformation of vectors in electromagnetic (left) and acoustic or compressional elastodynamics (right). The white converging arrows denote which component of each vector is compressed by the coordinate transformations (From Cummer et al. 30)

the direction of the vector component. Recall that our qualitative discussion above, summarized in Fig. 13.6, showed that this is precisely how the velocity vector must transform in a compressed wave in order for transformation acoustics to work. Note that the elements of the volume vector $\left[v^1 v^2 v^3\right]^{\mathrm{T}}$ are the contravariant components of $\vec{v}$ in the nonorthogonal coordinate system while the element of the vector $\vec{v}$ is the component in the original orthogonal coordinate system.

Multiplying (2) (with $\lambda(\vec{r}) = 1$) by $Q_1 Q_2 Q_3 V_{\text{frac}}$ and using (13.23) results in the equation in the transformed coordinates,

$$i\omega p = \kappa(\vec{q})\kappa \nabla_q \cdot \tilde{\vec{v}} \tag{13.25}$$

with

$$\kappa(\vec{q}) = (Q_1 Q_2 Q_3 V_{\text{frac}})^{-1} \tag{13.26}$$

This demonstrates the coordinate to function invariant of (13.17) provided that the bulk modulus is modified according to (13.26) and the velocity vector is transformed according to (13.25). More generally, this also shows how a vector must transform in order for the gradient operator to maintain its basis form.

Cummer et al. [30] derived how (13.16) and therefore the gradient operator transforms under a coordinate change using the gradient theorem and integrating ∇p along a short length in the q_1 coordinate directions, they find that

$$\nabla p \cdot Q_1 \hat{u}_1 = \frac{\partial p}{\partial q_1} = \left(\nabla_q p\right)^1 \tag{13.27}$$

The left-hand side contains the scaled covariant components of ∇p which must be converted to covariant components before it can be equated componentwise to $\nabla_q p$, the gradient in the transformed coordinates. They find that

$$\nabla_q p = \overline{\overline{Q}}_{\text{par}} \bar{\bar{h}}^{-1} (\nabla p) \tag{13.28}$$

where $\overline{\overline{Q}}_{\text{par}}$ is the diagonal tensor containing coordinate scaling factors parallel to the direction of the vector component or

$$\overline{\overline{Q}}_{\text{par}} = \begin{bmatrix} Q_1 & 0 & 0 \\ 0 & Q_2 & 0 \\ 0 & 0 & Q_3 \end{bmatrix} \tag{13.29}$$

and

$$\bar{\bar{h}}^{-1} = \begin{bmatrix} \hat{u}_1 \cdot \hat{u}_1 & \hat{u}_1 \cdot \hat{u}_2 & \hat{u}_1 \cdot \hat{u}_3 \\ \hat{u}_2 \cdot \hat{u}_1 & \hat{u}_2 \cdot \hat{u}_2 & \hat{u}_2 \cdot \hat{u}_3 \\ \hat{u}_3 \cdot \hat{u}_1 & \hat{u}_3 \cdot \hat{u}_2 & \hat{u}_3 \cdot \hat{u}_3 \end{bmatrix} \tag{13.30}$$

Note that this $\bar{\bar{h}}^{-1}$ is the same as $\bar{\bar{g}}^{-1}$ defined by Pendry et al. [22]. They rename this tensor because they will use $\bar{\bar{g}}$ later to denote the metric tensor which is not quite the same as this $\bar{\bar{h}}$.

Finally multiplying (1) (with $\rho(\vec{r}) = 1$) by $\overline{\overline{Q}}_{par}$, they find

$$p_q p = i\omega \overline{\overline{Q}}_{\text{par}} \bar{\bar{h}}^{-1} \rho_0 \vec{v} = i\omega \overline{\overline{Q}}_{\text{par}} \bar{\bar{h}}^{-1} \overline{\overline{Q}}_{\text{par}}^{-1} V_{\text{frac}}^{-1} \rho_0 \vec{v} \tag{13.31}$$

leaving us with the equivalent of (1) in fully transformed coordinates

$$\nabla_q p = i\omega \bar{\bar{\rho}} \rho_0 \vec{v} \tag{13.32}$$

with

$$\bar{\bar{\rho}} = \overline{\overline{Q}}_{\text{par}} \bar{\bar{h}}^{-1} \overline{\overline{Q}}_{\text{par}}^{-1} \overline{\overline{V}}_{\text{frac}}^{-1} \tag{13.33}$$

Equations (13.11) and (13.18) show that the acoustic equations are fully transformation invariant with the modified material parameters in (14.26) and (14.33).

They further show that these experiments are equivalent to those shown by Chen and Chan [24] purely by analogy with electromagnetics through the electric conductivity equation [31] and those derived by Greenleaf et al. [26] for the general scale Helmholtz equation. Consequently, cloaking shell, concentrator and other devices that have been designed theoretically by electromagnetic can also be realized for acoustics provided that the bulk modulus and anisotropic effective mass density tensor can be realized in practice as specified by (13.26) and (13.33). This first principles derivation without using analogy shows explicitly in (13.23) how the acoustic velocity vector must transform under coordinate change, which as noted above is different from how the $\vec{E}$ and $\vec{H}$ field, transform in electromagnetics. The scalar pressure is however, not changed by the coordinate transformations and thus like phase fronts and power flow lines, is simply deformed by any coordinate transformations.

C. Application to a Specific Example

We consider the spherical cloaking transformations [6] as illustrated in Fig. 13.7 and specified by $r' = a + r(b-a)/b$ where a and b are constants and $b > a$. This coordinate transformations is orthogonal, and then $\bar{\bar{h}} = 1$ and $V_{\text{frac}} = 1$ which are good simplification. The Q_i length scaling factors are straightforward to calculate provided one realizes that the azimuthal and polar angles and not length, as in Cartesian coordinates and (13.17) must be modified slightly. The Q_i are defined by the ratio of infinitesimal lengths in the transformed and untransformed coordinates and thus,

$$Q_r = \frac{\mathrm{d}r}{\mathrm{d}r'} = \frac{b}{b-a}, \quad Q_\phi = \frac{r\mathrm{d}\emptyset}{r'\mathrm{d}\phi'} = \frac{b}{b-a}\frac{r'-a}{r'} \tag{13.34}$$

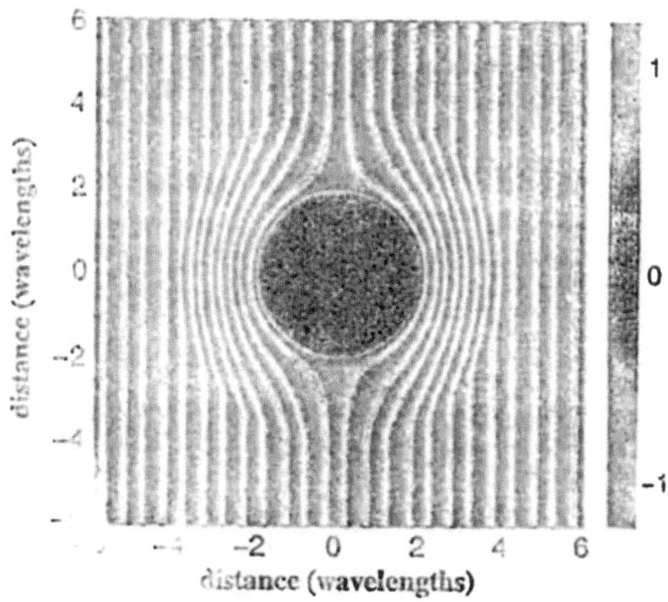

Fig. 13.7 (Colour online) The real part of the pressure field in the $r-\theta$ plane of the problem domain is computed from the series solution. The plane wave is incident from the left (From Cummer et al. 2008)

$$Q_\Theta = \frac{r \sin\theta}{r' \sin\theta'} \frac{d\theta}{d\theta'} = \theta_\varphi \tag{13.35}$$

in agreement with the parameter found through other approaches by Chen and Chan [25], Greenleaf et al. [26] and Cummer et al. [32].

Thus, Cummer et al. [30] showed $\vec{E}$ and $\vec{H}$ of electromagnetics transform differently from $\vec{v}$ of acoustic waves under coordinate transformations. It shows that a first principle analysis of the acoustic equation under arbitrary coordinate transformations confirms that the divergence operator is preserved only if velocity transform in this physically correct way.

13.5 Acoustic Metamaterial with Simultaneous Negative Mass Density and Negative Bulk Modulus

This is a different approach from that of using multiple scattering theory (MST) to produce acoustic negative refraction using phonons crystals [17, 33] (of Sect. 13.3 of MST) and also different from that of the fabrication of acoustic metamaterials for acoustical cloaking based on the invariance of the acoustic field equations under coordinate transformations. The concept is based on the gauge invariance of the acoustic field equations [5]. That is, there is no change in the form of the acoustic field equation with the replacement of the density and bulk modulus by negative density and negative bulk modulus. We have shown in section two that the concept of negative permittivity and negative permeability giving rise to negative refraction ([1] of section one) can also be explained by the gauge invariance of the Maxwell's equation with the replacement of the positive permeability and positive permittivity by negative permeability and negative permeability. In fact, gauge invariance is more appropriate than the approach of Veselago [1] using the dispersive relation as the starting point to introduce negative permeability and negative permittivity as this will give rise to the restriction that only single frequency electromagnetic cloaking is allowed and also the Veselago's [1] dispersion relation is used only to the isotropic case whereby most acoustic cloaking materials are anisotropic.

Applying gauge invariance of acoustic fields [34] to negative refraction, broadband double-negative spectral range in the structure can be obtained [28]. This is also an experimental verification of my hypothesis on the gauge invariance of acoustic fields [5]. Lee et al. [28] fabricated an acoustic double negativity (DNG) acoustic metamaterial with both membranes and side holes (Fig. 13.8). Here the acoustic waves are governed by Eqs. (13.36) and (13.37)

$$-\nabla p = \left[\rho - \frac{\kappa}{\omega^2}\right] \frac{\partial \vec{u}}{\partial A} \tag{13.36}$$

and

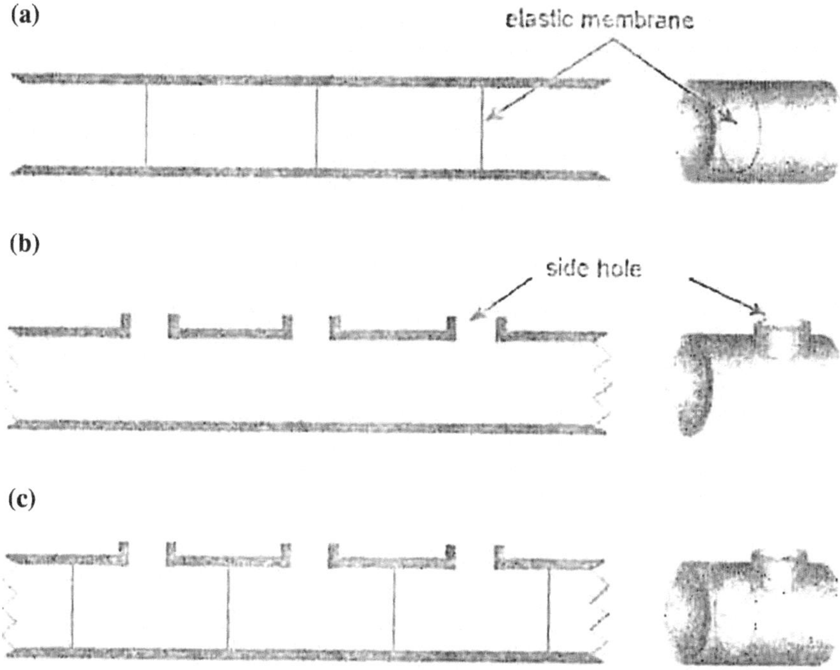

Fig. 13.8 **a** One-dimensional SAE structure consisting of thin tensioned elastic membranes in a tube. Negative effective density is observed in this system. **b** A tube with an array of side holes that exhibits negative effective modulus. **c** An acoustic DNG structure with both membranes and side holes (From Lee et al. [28])

$$\nabla \cdot \vec{u} = -\left[\frac{1}{B} - \frac{\sigma_{SH}^2}{\rho_{SH} A\omega^2}\right]\frac{\partial p}{\partial A} \tag{13.37}$$

where

κ new elastic modulus,
$\vec{u}$ velocity of the fluid (air in this case),
ρ dynamic mass density,
B bulk modulus,
A cross section of the tube,
σ_{SH} SH-cross sectional-density,
ρ_{SH} SH-mass-density.

The existence of the side holes (SH) does not modify Eq. (1). Likewise, because the membranes do not sink any fluid, Eq. (2) is still valid. Then, the system is described by the dynamic and continuity equations

$$-\nabla p = \rho_{\text{eff}}\left(\frac{\partial y}{\partial A}\right) \nabla \cdot \vec{u} = -\left(\frac{1}{B_{\text{eff}}}\right)\left(\frac{\partial p}{\partial A}\right)$$

with the effective density and modulus given by (14.38) and (14.39)

$$P_{\text{eff}} = \rho' - \frac{\kappa}{\omega^2} = \rho'\left[1 - \frac{\omega_{\text{SAE}}^2}{\omega^2}\right] \tag{13.38}$$

$$B_{\text{eff}} = \left[\frac{1}{B} - \frac{\sigma_{\text{SH}}^2}{\rho_{\text{SH}} A\omega^2}\right]^{-1} = B\left[1 - \frac{\omega_{\text{SH}}^2}{\omega^2}\right]^{-1}$$

where

$$\omega_{\text{SAE}} = \text{critical frequency} = \sqrt{\frac{\kappa}{\rho'}} \tag{13.39}$$

The resulting wave equation gives the phase velocity,

$$v_{\text{ph}} = \pm\sqrt{\frac{B_{\text{eff}}}{\rho_{\text{eff}}}} = \pm\sqrt{\frac{B}{\rho'\left(1 - \omega_{\text{SAE}}^2/\omega^2\right)\left(1 - \omega_{\text{SH}}^2/\omega^2\right)}}$$

where

$$\omega_{\text{SH}} = \left(B\sigma_{\text{SH}}^2/A\rho_{\text{SH}}\right)^{1/2} \tag{13.40}$$

The experimental setup is given in Fig. 13.9a.

It consists of a nonmetal tube on the left and the DNG metamaterial on the right. The absorbers at both ends completely absorb the acoustic energy, preventing any reflection so the system behaves as if it extends to infinity. This eliminates concerning about the effect of the finite number of cells used in the experiment, as well as the interference effect from the reflected waves. The sound source rejects acoustic energy into the tube through a small hole, generating incident waves propagating to the right. At the boundary, a position of the incident energy is reflected and the rest is transmitted into the metamaterial regions. On the metamaterial side, the transmitted acoustic energy flows steadily to the right until it hits the absorber.

Pressure was measured as a function of time and position on both the normal tube side and the metamaterial side. It can be seen that on the normal tube side, the wave proceeds forward, but on the metamaterial side, the wave propagated as indicated by the arrows. Clearly, the wave on the metamaterial propagated in a direction antiparallel to the energy flow. This confirms the theoretical prediction of negative phase velocity. It was noted that the amplitudes and the apparent phase velocity in the normal tube are deviated from the actual values of the incident wave because of the interference of the reflected wave from the boundary. In the metamaterial, there is no such interference effect because there is no reflected wave.

The comparison between the theory and experiment is shown in Fig. 13.10.

Theoretically expected single negative gap is experimentally confirmed by the transmission data (inset). In the DNG and DPS (double positions) pass bands, the phase velocities experimentally determined agree well with the theoretical values.

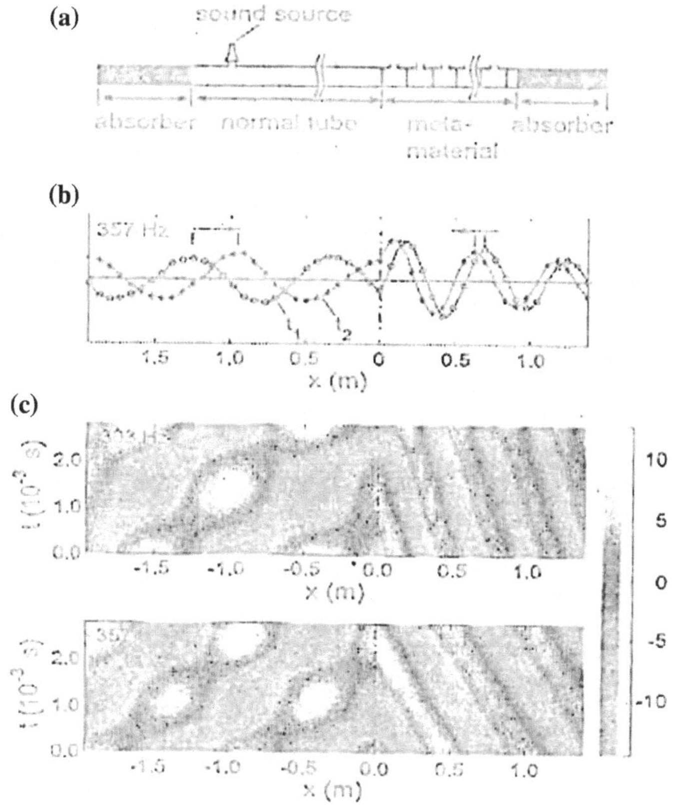

Fig. 13.9 **a** Experimental setup for the transmission and phase velocity measurements. **b** "Snapshots" of measured pressure distribution showing backward wave propagation in the metamaterial ($x > 0$). **c** Characteristic diagrams of pressure measurements for the frequencies 303 and 357 Hz. Negative slopes of the wave-paths in the metamaterial sides ($x > 0$) indicate negative phase velocities (From Lee et al. [28])

The calculations given as accurate description of the behaviour of the phase velocity in the frequency range from 250 to 1500 Hz, which is broadband. Because the experiment confirms the theoretical prediction of negative phase velocity, it can be concluded that the density and the bulk modulus actually become simultaneously negative in the frequency range below 440 Hz.

We would like to point out that the novel concept of spatially anchored elasticity [16] was used. This uses a homogenized structure of membranes to produce negative effective density. This is termed spatially anchored elasticity (SAE) because the fluid is elastically anchored in space by the membranes. The new elasticity can be regarded as an intrinsic variables that characterizing the behaviour of the metamaterial according to Eq. (13.41)

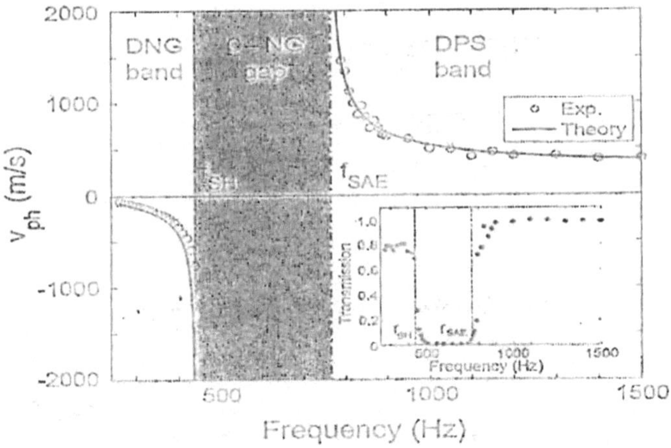

Fig. 13.10 Transmission (inset) and phase velocities of the present acoustic DNG medium (From Lee et al. [28])

$$\nabla p = -\kappa \vec{\xi} \tag{13.41}$$

where

κ new elastic modulus,
ξ displacement of the fluid,
p pressure of the fluid.

Furthermore, by making additional side holes along the tube wall, acoustic DNG materials were obtained and backward wave propagation was observed. The constructed structure exhibited DNG characteristic in the spectral range from 240 to 440 Hz which is broadband unlike the electromagnetic case which is limited only to a single frequency due to dispersion. The phase velocity in this band was negative and highly dispersive.

Again this proves that acoustic metamaterial cannot be just transplanted by analogy from the electromagnetic case. It has to be based on the theory of elasticity unlike for the electromagnetic which is based on the dispersion relation of Veselago [1].

13.6 Acoustical Cloaking Based on Nonlinear Coordinate Transformations

So far the coordinate transformations used in acoustical cloaking are based on linear coordinate transformations [6]. Akl et al. [35] extended to nonlinear transformation using

$$r^1 = a + (b - a)\left(\frac{r}{b}\right)^n \tag{13.42}$$

where n = an arbitrary transformation exponent that accounts for the degree of nonlinearity in the transformation and can be used as an additional degree of freedom in designing and controlling the bending of the acoustic wave inside the cloak. For unity value of n, the transformation returns back to the linear transformation proposed by Cummer and Schurig [6].

Linear transformation is effective for the case of rigid objects. However, the cloaking becomes less perfect and dependent on the selection domain when flexible objects are considered where its permeable nature might induce considerable absorption of sound wave which would bring less perfection to the cloak. Akl et al. [35] have presented acoustic cloaking based on different nonlinear coordinate transformations. They developed a finite element model developed through time-harmonic analysis to study the preserve field distribution using different nonlinear coordinate transformations. Such transformations have shown considerable improvements to the cloak performance when applied to flexible objects allowing for wider applicability bandwidth (broadband) as well as for providing additional control of the shape of acoustic wave bending inside the cloak region.

For a metamaterial anisotropic acoustic cloak of a flexible object, the cloak works in a limited frequency range around its resonant frequency. In order to show this fact, a quantifiable measure for the cloak's performance has been developed. They proposed a new performance indicator of the cloak's quality using the acoustic pressure value at a preselected set of points downstream of the cloak. The set of points selected for pressure measurements were distributed along the fluid in such a way to accurately predict the calculation from ideal cloak. The proposed indicator is based on the fact that for ideal cloak, the r.m.s. of the difference between the acoustic pressure values along the wave front downstream of the cloak and a reference value measured along the same wavefront at a reference wave propagation line tend to be zero. A reference wave propagation line, located at the middle of the domain is quite a good unbiased choice. This process is repeated with as much planes in the axial direction (along wave propagation lines) and lateral directions as needed to scan the entire fluid domain, where the sum of all the calculated values is divided by the number of measurement points as given in Eq. (13.43)

$$\text{P.I.} = \sum_{j=1}^{j\max} \sqrt{\sum_{i=1}^{i\max} \left(P_{i,j} - P_{i,\text{ref}}\right)^2} \tag{13.43}$$

to yield the targeted performance indicator (P.I.). The measurement grid points are as illustrated in Fig. 13.11.

In this case, any determination of the cloak performance would result in a positive r.m.s. value of the proposed pressure difference. The number of points selected was large enough to capture even the smallest deviation from the ideal cloak performance.

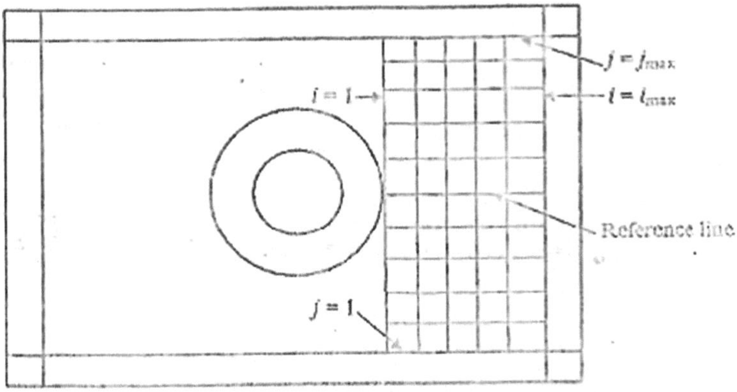

Fig. 13.11 Schematic of the measurement grid points required revaluation the performance indicator (From Akl et al. [35])

In Eq. (13.2), i is the measurement point index along the wave propagation line (axial direction), which j represents the point index in the lateral directions. Based on this indicator, the performance of an anisotropic acoustic cloak surrounding a flexible object is quantified at different excitation frequency values such that the larger the indicator values, the more deterioration in the cloak performance is.

The proposed nonlinear transformations of Akl et al. [35] have proven to improve the way in which the acoustic metamaterial anisotropic cloak works away from the limited frequency values. This is shown in Fig. 13.12 by plotting the performance indicator of the linear cloak and one of the nonlinear cloaking over the frequency range under study. It is clear that the nonlinear cloak performance at frequency ranges away from these domain resonant frequencies is much better. The same conclusion is drawn from Fig. 13.13 where the acoustic pressure field for both the linear and nonlinear transformations for frequency values away from the domain resonance frequency is plotted. In order to show the degree by which the nonlinear transformations has improved the cloaking performance, the difference between the performance indicator values for nonlinear transformations with minimum PI value and those for linear transformations at each frequency are calculated and plotted against the excitation frequency as shown in Fig. 13.14. In this figure, the higher the positive difference, the more improvement of the acoustic cloak performance is achieved. Once more, it is evident that a perfect linear acoustic cloak is achievable only at the same specific frequency values and the proposed nonlinear transformations have improved significantly the way in which the acoustic metamaterial anisotropic cloak works away from the limited frequency values. Although the simulations of the acoustic metamaterial anisotropic cloak around flexible objects encounters some sort of numerical error, the proposed nonlinear transformations opens the door for searching for different coordinate transformations function that would lead to simulation results insensitive to the solution domain dimensions.

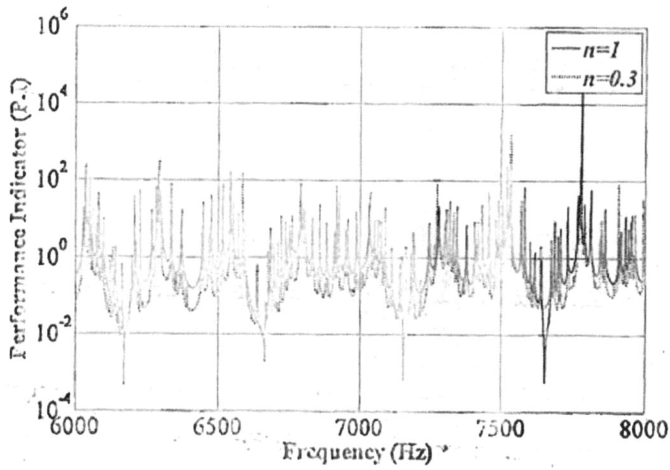

Fig. 13.12 Nonlinear ($n = 0.3$) acoustic cloak performance against the linear cloak when surrounding the host medium at different frequency values (From Akl et al. [35])

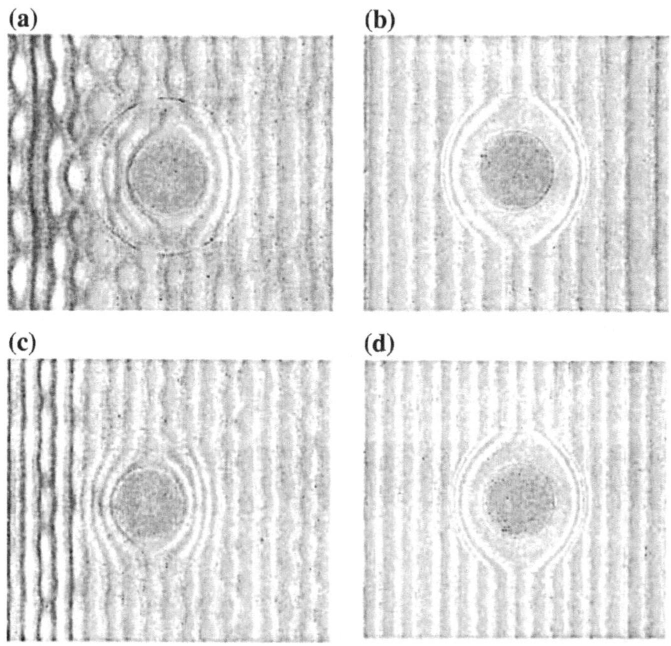

Fig. 13.13 Full wave time-harmonic acoustic pressure field plot of the analysed ideal cloak with water as base medium: **a** linear at 6000 Hz, **b** nonlinear ($n = 0.3$) at 6000 Hz, **c** linear at 7000 Hz and **d** nonlinear ($n = 0.3$) at 7000 Hz (From Akl et al. [35])

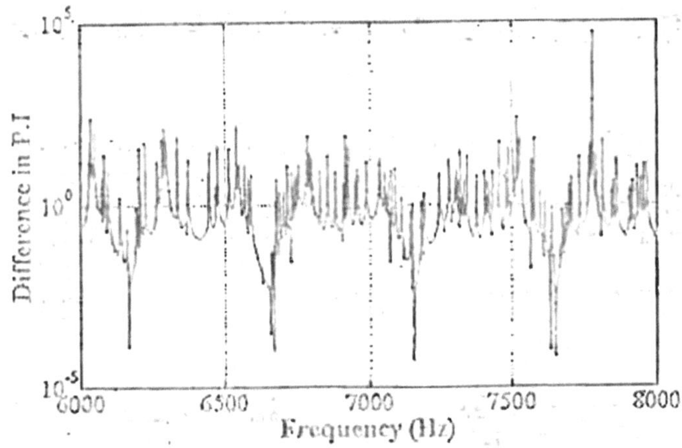

Fig. 13.14 Performance indicator difference between linear and nonlinear cloaks at different frequency values (From Akl et al. [35])

13.7 Acoustical Cloaking of Underwater Objects

A group at the Mechanical Engineering Department of the University of Illinois led by Nicholas Fang has created a numerical model to build a metamaterial cloak that guides sound waves around objects in water. The model is based on the acoustic lumped circuit network. The unit cell of the network is so small compared with to the wavelength of the sound that it becomes and effective anisotropic medium that guides sound flow around the cloaked object. Computer simulations demonstrated that the numerical model successfully achieved a cloaking effect. The next step is to construct and test an actual physical version of the cloak based on that numerical model. If the metamaterial cloak also works, considerably more works need to be done before the cloak could be scaled up to hide a ship or a submarine. Their mesh model is based on cloaking an object with a diameter of about 0.67 times the wavelength of light—a far cry from the 50-foot beam of a nuclear submarine. Their work is published in the May 15 2009 issue of the Physical Review Letters [36].

13.8 Extension of Double Negativity to Nonlinear Acoustics

The lossless form of nonlinear acoustic wave equation up to the third-order elastic coefficient can be given by Thurstone and Shapiro [37] as

$$\ddot{u} = \frac{M_2}{\rho_0}\frac{\partial^2 u}{\partial x^2} + \frac{M_3}{\rho_0}\left(\frac{\partial^2 u}{\partial x^2}\right)\left(\frac{\partial u}{\partial x}\right) \tag{13.44}$$

where

u displacement,
x Lagrange coordinate in the direction of motion of a particle, and anisotropic solid is used.

where M_2 is a linear combination of second-order elastic coefficient and M_3 is a linear combination of second and third-order elastic coefficient.

To allow for energy dissipation, Eq. (13.44) is modified by adding a term to include the frequency dependent attenuation coefficient $\alpha = \alpha(\omega)$, to the right-hand side

$$\ddot{u} = \frac{M_2}{\rho}\frac{\partial^2 u}{\partial x^2} + \frac{M_3}{\rho}\left(\frac{\partial^2 u}{\partial x^2}\right)\left(\frac{\partial u}{\partial x}\right) + \frac{2\alpha}{\omega^2}C^3\frac{\partial^2 u}{\partial x^2 \partial t} \tag{13.45}$$

where

$C^2 = M_2/\rho$ speed of propagation of an infinitesimal amplitude sound wave and
ρ mass density of medium.

By replacing ρ by $-\rho$ and M_2 and M_3 by $-M_2$ and $-M_3$. There is no change in the form of equation. Hence the nonlinear acoustic new equation is also gauge invariant in the mass density and the elastic coefficient.

References

1. Veselago, V.G.: The electrodynamics of substances with simultaneous negative values of ε and μ. Sov. Phys. Usp. **10**(4), 509–514 (1968)
2. Mandel'stam, L.I.: JETP **15**, 475 (1945)
3. Pendry, J.B., Holden, A.J., Robbins, D.J., Stewart, W.J.: Magnetism from conductors and enhanced non-linear phenomena. IEEE Trans. Microwave Theory Technol. **47**(11), 2075–2984 (1999)
4. Shelby, R.A., Smith, D.R., Schultz, S.: Experimental evidence of a negative index of refraction. Science **292**(5514), 77 (2001)
5. Gan, W.S.: Gauge invariance approach to acoustic fields. In: Akiyama, I. (ed) Acoustical Imaging, vol. 29, pp. 389–314. Springer (2007)
6. Cummer,S.A., Schurig, D.: One path to acoustic cloaking, New J. Phys. **9**, 45 (2007)
7. Korringa,J.: Physica (Amsterdam) **XIII**, 392 (1947)
8. Kohn, W., Rostoker, N.: Phys. Rev. **94**, 1111 (1951)
9. Liu, Z., Chan, C.T., Sheng, P., Goertzen, A.L., Page, J.H.: Elastic wave scattering by periodic structures of spherical objects: theory and experiment. Phys. Rev. B **62**(4), 2446–2457 (2000)
10. Sigalas,M.M., Economou, E.N. J. Sound Vib. **158**, 377 (1992)
11. Kushwaha, M.S., Halevi, P., Dobrzynski, L., Djafrari-Rouhani, B.: Phys. Rev. Lett. **71**, 2022 (1993)
12. Sanchez-Perez, J.V., Caballero, D., Martinez-Sala, R., Rubio, C., Sanchez-Dehesa, J., Meseguer, F.: Phys. Rev. Lett. **80**, 5325 (1998)
13. Kafesaki, M., Economou, E.N.: Phys. Rev. B **52**, 13317 (1995)
14. Yang, S., et al.: Phys. Rev. Lett. **88**, 104301 (2002)
15. Wolfe, J.P.: Imaging Phonons: Acoustic Wave Propagation in Solids. Cambridge University Press, Cambridge, England (1998)

16. Liu, Z., et al.: Phys. Rev. B **62**, 2446 (2000)
17. Zhang, X., Liu, Z.: Negative refraction of acoustic waves in two-dimensional phononic crystals. Appl. Phys. Lett. **85**(2), 341–343 (2004)
18. Luo, C., Johnson, S.G., Joannopuolos, J.D., Pendry, J.B.: Phys. Rev. B **65**, 201104 (2002)
19. Luo, C., Johnson, S.G., Joannopoulos, J.D.: Appl. Phys. Lett. **83**, 2352 (2002)
20. Lai, Y., Zhang, X., Zhang, Z.Q.: Appl. Phys. Lett. **79**, 3224 (2001)
21. Pendry, J.B.: Phys. Rev. Lett. **85**, 3966 (2000)
22. Pendry, J.B., Schurig, D., Smith, D.R.: Controlling electromagnetic fields. Science **312**, 1780–1782 (2006)
23. Schurig,D., Mock, J.J., Justice, B.J., Cummer, S.A., Pendry, J.B., Starr, A.F., Smith D.R.: Metamaterial electromagnetic cloak at microwave frequencies. Science **314**, 977–980 (2006)
24. Milton, G.W., Briane, M., Willis, J.R.: On cloaking for elasticity and physical equations with a transformation invariant form. New J. Phys. **8**, 248 (2006)
25. Chen, H., Chan, C.T.: Acoustic cloaking in three dimensions using acoustic metamaterials. Appl. Phys. Lett. **91**, 183518 (2007)
26. Greenleaf, A., Kurylev, Y., Lassas, M., Uhlmann, G.: Comment on "Scattering Derivation of a 3D Acoustic Cloaking Shell" (2008)
27. Cummer, S.A., Raleigh, M., Schurig, D.: New J. Phys. **10**, 115025–115034 (2008)
28. Lee, S.H., Kim, C.K., Park, C.M., Seo, Y.M., Wang, Z.G.: Composite acoustic medium with simultaneously negative density and modulus. In: Proceedings of ICSV17 (2010)
29. Cheng, Y., Xu, J.Y., Liu, X.J.: One-dimensional structured ultrasonic metamaterials with simultaneously negative dynamic density and modulus. Phys. Rev. B **77**, 045134 (2008)
30. Cummer, S.A., Rahm, M., Schurig, D.: Material parameters and vector scaling in transformationacoustics. New J. Phys. **10**, 115025 (2008)
31. Greenleaf, A.et al.: Anistropic conductivities that cannot ne detected by EIT. Physiol. Meas. **24**, 413–419 (2003)
32. Cummer, S.A., et al.: Scattering theory derivation of a 3D acoustic cloaking shell. Phys. Rev. Lett. **100**, 024301 (2008)
33. Yang, S., Page, J.H., Liu, Z., Cowan, M.L., Chan, C.T., Sheng, P.: Focusing of sound in a 3D phononic crystal. Phys. Rev. Lett. **93**(2), 024301-1–024301-4 (2004)
34. Hu, J., Zhou, X., Hu, G.: A numerical method for designing acoustic cloak with arbitrary shapes. Comput. Mater. Sci. (Elsevier) **46**, 708–712 (2009)
35. Akl, W., Elnady, T., Elsabbagh, A.: Improving acoustic cloak bandwidth using nonlinear coordinate transformations. In: Proceedings of ICSV17 (2010)
36. Fang, N., Zhang, S.: Phys. Rev. Lett. (2009)
37. Thurston, R.N., Shapiro, M.J.: J. Acoust. Soc. Am. **41**, 1112 (1967)

Chapter 14
New Acoustics Based on Metamaterials

Abstract An introduction is followed by topics on new acoustics and acoustical imaging. The background of phononic crystals is given. The theory of phononic crystals—the multiple scattering theory is provided with details of calculation and discussion of results. An alternative theory of negative refraction based on gauge invariance (coordinate transformations) is followed with the following coverage: gauge invariance as a unified theory of negative refraction and cloaking, generalized form of Snell's law for curvilinear coordinates, design of a perfect lens using coordinate transformations and a general cloaking lens. The reflection and transmission of sound wave at interface of two media with different parities are given. The theory of diffraction by negative inclusion is stated with the following coverage: formulation of forward problem of diffraction tomography, modelling diffraction procedure in a negative medium, results of numerical simulation and points to take care of during numerical simulation. Then, there is the extension to theory of diffraction by inclusion of a general form of mass density and bulk modulus manipulated by predetermined direction of sound propagation. A new approach to diffraction theory—a rigorous theory based on material parameters is described. Negative diffraction derived from reflection invariance (right–left symmetry)—a new approach to negative refraction is covered. A unified theory for the symmetry of acoustics fields in isotropic solids, negative refraction and cloaking is provided. There is application of new acoustics to acoustic waveguide and new elasticity. Also, there is nonlinear acoustics based on the metamaterials covering properties and nonlinear acoustic metamaterials for sound attenuation application. There is ultrasound attenuation in acoustic metamaterials covering mechanism of energy transfer and wave attenuation and applications. Application of phononic crystal devices is provided. A comparison of the significance of the role played by gauge theory and multiple scattering in metamaterial—a sum-up of the theories of metamaterial is given. Finally, conclusions are provided.

14.1 Introduction

Metamaterials are composite materials with artificial properties defined by their subwavelength structures rather than their chemical composition. They also must have periodic structures. There are many types of metamaterials such as bandgap

W. S. Gan, *Gauge Invariance Approach to Acoustic Fields*,
https://doi.org/10.1007/978-981-13-8751-7_14

(BNG) metamaterials, double-negative (DNG) metamaterial and single-negative (SNG) metamaterial. Examples of bandgap metamaterial are photonic crystals and phononic crystals. The world's oldest metamaterials are photonic crystals. Photonic crystals have been studied in one form or another since 1887. Lord Rayleigh started his study in 1887 by showing that such systems have a one-dimensional photonic bandgap, spectral range of large reflectivity, known as stopband. Today, such structures are used in a diverse range of applications: from reflective coatings to enhancing the efficiency of LEDs to highly reflective mirrors in certain laser cavities. Double-negative metamaterials are those with negative permittivity and negative permeability for the electromagnetic case and negative mass density and negative bulk modulus for the acoustic case. In single-negative metamaterials, either permittivity or permeability or either mass density or bulk modulus is negative. Metamaterials have unique properties and anomalous behaviours that give rise to new physics. For instance, although light consists of an electric field and a magnetic field, ordinary optical materials, such as optical microscope lenses, have a strong reaction only to the electric field. The corresponding magnetic interaction is essentially nil. This results in only the most common optical effects, such as ordinary refraction with common diffraction limitations in lenses and imaging. Since the beginning of optical sciences, the ability to control light with materials has been limited only to these common effects. Electromagnetic metamaterials, on the other hand, are capable of a very strong interaction or coupling with the magnetic component of light. Therefore, the scope of response to radiated light is expanded beyond the ordinary optical limitations that are described by the sciences of physical optics and optical physics. In addition, as artificially constructed materials, both the magnetic and the electric components of the radiated light can be controlled at will, in any desired fashion as it travels, or more accurately propagates, through the material. This is because a metamaterial's behaviour is typically formed from individual components and each component responds independently to a radiated spectrum of light. In metamaterials, one can craft the nanoscale spatial distribution with the future potential of choosing the refractive index in almost any desired way and thus moulding the flow of light with subwavelength precision.

Double-negative electromagnetic metamaterial was first proposed theoretically by Veselago [1] in 1967. Pendry [2] was the first to theorize a practical way to make a double-negative electromagnetic metamaterial which is also known as the left-handed material because the electric field vector, the magnetic field vector and the vector of wave propagation form a left-handed triplet. Pendry's initial idea was that metallic wires aligned along the direction of propagation could provide a metamaterial with negative permittivity ($\varepsilon < 0$). Note however that, natural materials such as ferroelectrics were already known to exist with negative permittivity, and the challenge was to construct a material which also showed negative permeability ($\mu < 0$). In 1999, Pendry [2] theorized that a split-ring resonator (SRR) with its axis placed along the direction of wave propagation could provide a negative permeability. In the same paper, he showed that a periodic array of wires and ring could give rise to a negative refractive index. Pendry's [2] theoretical proposal was experimentally demonstrated by Smith et al. [3] in 2000. Acoustic waves, on the other hand, are of a mechanical nature and do not possess those magnetic properties as electromagnetic waves. Hence, the physical mechanism of achieving double negativity for acoustic waves is different from that of electromagnetic waves. Although for acoustic waves the double negativity is also achieved by resonance structure, they are from a single resonance structure as distinct from Pendry's [2] case of electromagnetism in which the negative permeability and negative permittivity originate from different resonance mechanisms; for instance, the negative permeability originates from the resonance structure of the split-ring resonator.

Double-negative acoustical metamaterial has been first theorized and numerically simulated by Li and Chan in 2004 [4]. Li et al. [4] used analogies between electromagnetic waves and sound waves. It can be shown from the first principles that the acoustic equation of motion is invariant in the form (or gauge invariant) to negative mass density and negative bulk modulus (or compressibility).

For left-handed metamaterials, the Poynting vector is in the opposite direction to wave propagation direction or the wave is backpropagating towards the source. Together with the negativity of the two material parameters, these two important properties give rise to new phenomena in refraction, diffraction and scattering of acoustic waves in the left-handed media. These three are the basic mechanisms of sound propagation in the medium. For each new mechanism, there are several new devices to be developed.

For instance, the capability to control and manipulate refraction gives rise to the new phenomenon of negative refraction and the design of the perfect lens defeating the Rayleigh resolution limit. Likewise, the ability to control and manipulate diffraction will give rise to a new theory of diffraction and a new form of diffraction and the ability to control and manipulate scattering will give rise to a new theory of scattering and a new form of scattering. Hence, we call this new field new acoustics.

14.2 New Acoustics and Acoustical Imaging

Acoustical imaging is the collection of the information of sound propagation through a medium (solids and fluids), and this information depends on the three basic mechanisms of sound propagation through a medium (refraction, diffraction and scattering). The capability of metamaterial being able to control and manipulate refraction, diffraction and scattering will give rise to the manipulation of acoustical imaging. Hence, new acoustics will certainly cause a revolution in the world of acoustical imaging. Normal acoustic lens for refraction accounts only for the straight or linear propagation path of the sound wave. We can design a special lens using acoustical metamaterial for the bending and controlling of the propagation path of sound to a direction of our choice with the control and manipulation of diffraction and scattering besides refraction. This special lens can be a rectangular slab.

14.3 Background of Phononic Crystals

Phononic crystals like photonic crystals belong to the class of metamaterials known as bandgap metamaterials. This class of metamaterials is as important as double-negative (DNG) metamaterials.

The 1980s has seen a tremendous research interest in the physical properties of artificial structures comprised of two or more materials which differ in certain properties. The 1980s began with enthusiasm for microstructures of reduced dimensionality [5] such as quantum heterostructures, quantum wires and quantum dots. In the late 1980s, there has been ever-increasing interest in macrostructures known as photonic crystals [6]. Typically, these are periodic arrays of two transparent dielectrics. The periodicity plays a crucial role in understanding the physical properties of both microstructures and macrostructures. One important aspect in photonic crystals is the formation of forbidden frequency bands in which electromagnetic (EM) modes, spontaneous emission and zero-point fluctuations are all absent [7]. In this chapter, we deal with phononic crystals which are elastic composites made up of two materials with different elastic properties. In analogy to the case of phononic crystals, the emphasis will be on the existence of complete phononic bandgaps for the elastic composites and their practical implications. The earliest papers on theoretical studies of phononic crystals are given in Refs. [8–12]. In Ref. [8], a periodic array of parallel rods of circular cross section embedded in a different background was considered. The intersection of these rods with perpendicular plane forms a two-dimensional lattice. Sigalas et al. [8] investigated only the transverse polarization mode with displacement $\vec{u}(\vec{r}, t)$ parallel to the cylinders (and perpendicular to the Bloch wave vectors). The computations performed for Ni(Al) alloy cylinders in Al(Ni) alloy background exhibited absolute bandgap extending throughout the Brillouin zone. Sigalas et al. [8] also considered the mixed (longitudinal transverse) polarization modes for which $\vec{u}(\vec{r}, t)$ and the Bloch wave vectors are in the plane perpendicular to the cylinders. They found that Au cylinders in Be matrix exhibit a narrow, but complete gap, shared by both polarizations. An elaborate theory of acoustic band structure in periodic composites of arbitrary dimensionality was presented in Ref. [9].

Analogous to the photonic crystals in the frequency range of a phononic bandgap, vibrations, sound and phonons would be forbidden. From a practical point of view, a complete phononic bandgap could be engineered to provide a vibrationless environment for high-precision mechanical devices in a given frequency range. Phononic crystals in the form of piezoelectric and pyroelectric composites have already found useful applications in transducers for pulse-echo medical ultrasonic imaging and for transmitting and receiving signals under water [11–15].

An adequate phononic crystal composite transducer requires a detailed understanding of elastic wave propagation in periodic structures, so that the characteristic dimensions of the transducer disc and the composite may be correctly chosen. For such understanding, the availability of band structures is essential.

For acoustic devices based on periodic composites, the width of the bandgaps and the midgap frequency (frequencies) play an essential role. For these elastic composites, the details of the gaps depend on the structure of the crystal lattice, on the average density and average elastic constant contrast C_{44} of the two constituent solids, on their density contrast and elastic constant contrast, on the filling fraction of one of the constituents and on the lattice constant. An elaborate study of such a complicated dependence is a major task.

Experimental studies on phononic crystals are given in Refs. [14–17], demonstrating the usefulness of acoustic waves in illustrating general features of wave propagation in inhomogeneous or random media.

14.4 Theory of Phononic Crystals—The Multiple Scattering Theory (MST)

The theory of phononic crystals will involve the calculations of acoustic wave propagation in periodic structures. The ultimate purpose is to find the existence of bandgap in the periodic structures which is analogous to the photonic bandgaps in photonic crystals, and the calculations have been performed using the plane wave (PW) method. PW is based on the expansion of the periodic coefficients in the wave equations in Fourier sums. Study using the PW methods shows that bandgap can exist under rather extreme conditions that concern mainly the elastic parameters (density and velocities) of the components of the composite, the volume fraction of one of the two components and the topology. However, the PW method is unable to describe composites with solid scatterers in a fluid.

Here, we shall follow the multiple scattering (MS) approach of Kafesaki and Economou [18] based on the well-known (in the band-structure electronic community) Korringa–Kohn–Rostoker (KKR) theory [19, 20].

Spherical scatterers embedded in a fluid host were considered by Kafesaki et al. [18]. They start from the acoustic wave equation in a periodic medium:

$$\lambda(\vec{r})\nabla\left[\frac{1}{\rho(\vec{r})}\nabla p(\vec{r})\right]+\omega^2 p(\vec{r})=0 \tag{14.1}$$

where p = acoustic pressure, $\rho(\vec{r})$ = mass density, ω = sound angular, $\lambda = \rho\left(c_l^2 - c_t^2\right)$ = Lame's coefficient of the medium, and c_l, c_t = longitudinal and transverse sound velocities, respectively.

Equation (14.1) can be rewritten as

$$\nabla^2 p(\vec{r})+\frac{\omega^2}{c_o}p(\vec{r})+\omega^2\left[\frac{1}{c^2(\vec{r})}-\frac{1}{c_o^2}\right]p(\vec{r})+\rho(\vec{r})\left[\nabla\frac{1}{\rho(\vec{r})}\right]\nabla p(\vec{r})=0 \tag{14.2}$$

which has the form

$$H_o(\vec{r})p(\vec{r})+u(\vec{r})p(\vec{r})=0 \tag{14.3}$$

where $H_o(\vec{r})p(\vec{r}) = 0$, $H_o(\vec{r}) = \nabla^2 + \frac{\omega^2}{c_o^2}$ represents the wave equation in the absence of scatterers and c_o is the wave velocity in the host material.

Equation (14.3) has the same form as the Schrodinger equation for the electron waves. This analogy shows that one can extend the KKR to the acoustic case. However, one has to beware of the important difference between the electronic case and the acoustic case. That is, the potential in the acoustic case has a δ function singularity at the surface of the scatterers due to the factor ∇p^{-1}. Thus, the contribution of the surface scattering to the volume integrals is not negligible as in the electronic case [20].

It can be shown that [21] in a periodic system Eq. (14.2) is equivalent to the following integral equation:

$$p(\vec{r}) = \int_u G(\vec{r} - \vec{r}')V(\vec{r})p(\vec{r}')\mathrm{d}\vec{r}' \tag{14.4}$$

Here, V = volume of a unit shell, and $V(\vec{r})$ = local potential.

The Green function $G(\vec{r} - \vec{r}')$ is given by

$$G(\vec{r} - \vec{r}') = \sum_n \mathrm{e}^{i\vec{k}\cdot\vec{R}_n} G_o\left(\vec{r} - \vec{r}' - \vec{R}_n\right) \tag{14.5}$$

G_0 is Green's function [22] for the homogeneous equation $H_o(\vec{r})p(\vec{r}) = 0$:

$$G_o(\vec{r} - \vec{r}') = -\frac{1}{4\pi}\frac{\mathrm{e}^{iK_o|\vec{r}-\vec{r}'|}}{|\vec{r} - \vec{r}'|}, \quad K_o = \frac{\omega}{c_o} \tag{14.6}$$

$V(\vec{r})$ is zero outside the unit shell centred at the origin of the coordinate system. It is related to u by $u(\vec{r}) = \sum_n V\left(\vec{r}' - \vec{R}_n\right)$ and the pressure field, $p(\vec{r})$. It obeys Bloch's condition, $p\left(\vec{r} + \vec{R}_n\right) = \mathrm{e}^{i\vec{k}\cdot\vec{R}_n} p(\vec{r})$.

Taking into account that for acoustic waves the local potential V is nonzero only inside and at the surface of the scatterers (Eq. 14.2), the integral over the unit shell in Eq. 14.4 is reduced to an integral over the volume of a scatterer ($r' \le r_s$, r_s is the scatterer radius).

$$\int_V \mathrm{d}\vec{r}' = \lim_{\epsilon \to 0+} \int_{\vec{r}' \le r_s + \epsilon} \mathrm{d}\vec{r}' \tag{14.7}$$

The limiting procedure in Eq. (14.7) ensures that we approach the surface of the sphere from the inside, including thus the surface singularity.

By noticing that for nonoverlapping spheres and $\vec{r}, \vec{r}'$ inside a unit shell centred at the origin of the coordinate system, the function G obeys the equation

$$\nabla^2 G(\vec{r} - \vec{r}') + K_o^2 G(\vec{r} - \vec{r}') = \delta(\vec{r} - \vec{r}') \tag{14.8}$$

and by the wave equation and the Gauss theorem, the volume integral in Eq. (14.7) can be transformed into a surface integral.

After some algebraic manipulations, one can find that

$$\lim_{r' \to r_s^+} \int_{s'} [p(\vec{r}')\nabla_{r'} G(\vec{r} - \vec{r}') - G(\vec{r} - \vec{r}')\nabla_{r'} p(\vec{r}') d\vec{s}' = \begin{cases} p(\vec{r}') & \text{for } r > r_s \\ 0 & \text{for } r < r_s \end{cases} \tag{14.9}$$

where s' is a spherical surface of radius r', centred at the origin of the coordinates. $r' \to r_s^+$ in the above limit denotes that we approach the sphere surface from the outside. This is a direct consequence of (14.7), and it is very important for the acoustic case as the integrated functions are not continuous across the surface. The acoustic pressure is continuous, but its derivative has a step function discontinuity. Thus, the side limits do not coincide. The solution of Eq. (14.9) for $r < r_s$ gives the eigenfrequencies of our periodic system for each Bloch's vector $\vec{k}$. To obtain this solution, we use the fact that both the functions $G(\vec{r} - \vec{r}')$ and $p(\vec{r}')$ can be expanded in spherical functions of $\vec{r}$ and $\vec{r}'$:

$$G(\vec{r} - \vec{r}') = \sum_{\ell m} \sum_{\ell' m'} \begin{bmatrix} A_{\ell m \ell' m'} \overline{j}_\ell(K_o r) \overline{j}_{\ell'}(K_o r') \\ + K_o j_\ell(K_o r) y_{\ell'}(K_o r') \delta_{\ell \ell'} \delta_{m m'} \end{bmatrix} x Y_{\ell m}(\vec{r}) Y^*_{\ell' m'}(\vec{r}') \tag{14.10}$$

for $r < r'$

$$\begin{aligned} p(\vec{r}')\big|_{r' \geq r_s} &= p^{\text{out}}(\vec{r}') \\ &= \sum_{\ell m} a_{\ell m} [j_\ell(K_o r') + t_\ell h_\ell(K_o r')] Y_{\ell m}(\vec{r}') \end{aligned} \tag{14.11}$$

and the second-kind spherical Bessel functions of order ℓ and $h_\ell = j_\ell + i y_\ell$.

Substituting Eqs. (14.10) and (14.11) into Eq. (14.9), we obtain the final multiple scattering equations:

$$\sum_{\ell' m'} [A_{\ell m \ell' m'} - K_o I_m(t_{\ell'}^{-1}) \delta_{\ell \ell'} \delta_{m m'}] a_{\ell' m'} = 0 \tag{14.12}$$

The coefficients $A_{\ell m \ell' m'}$ in the above equations are called structure constants, and they depend on $\vec{K}$, ω and the lattice constants. The coefficients t_ℓ, relating the incident to the scattered field at each scatterer, can be calculated by solving a single scattering problem.

Equation (14.12) can be rewritten as

$$\sum_{\ell' m'} A_{\ell m \ell' m'} a_{\ell' m'} = 0 \Leftrightarrow \sum_{L'} \Lambda_{LL'} a_{L'} = 0, \quad L \equiv (\ell, m) \tag{14.13}$$

which corresponds to a linear homogeneous algebraic system. The condition for this system to have nonvanishing solutions, $\det(\Lambda) = 0$, gives the eigenfrequencies of the periodic composite.

A careful analysis of the above equations shows that the elastic parameters of the scattering material affect the calculation only through the scattering coefficients t_ℓ. t_ℓ can be calculated very easily and accurately for both solid and fluid scatterers. Thus, the method can be applied to both solid and fluid scatterers changing only the form of a single scattering problem. This, however, is not the only advantage of the method. Its most important advantage is that it can be applied also in the disordered system. It can treat system with positional as well as substitutional disorder.

A. Details of Calculation

The eigenmodes of a periodic system are obtained by requiring nonvanishing solutions for the linear homogeneous system (Eq. 14.13). Thus, one has to calculate the matrix X, the determinant of which has to be set equal to zero. The order of the matrix Λ depends on the number of the angular momentum term that we keep in the field function (14.11). In the calculations of Kafesaki and Economou [18], they obtained good convergence by keeping the maximum number of $\ell = \ell_{\max} = 3$ or 4 while for the lower bands they had good convergence with $\ell_{\max}$ less than 3.

Another parameter of the problem is the size of the periodic system. Kafesaki and Economou [18] have considered a system of 400–500 lattice vectors in the direct as well as the reciprocal lattice with excellent convergence.

Among the calculational problems of the multiple scattering method, one worth mentioning is the problem of the spurious "roots", that is, sign changes of the determinant that do not correspond to actual eigenfrequencies of the system.

B. Discussion of Results

Figure 14.1 shows the band structure along the LΓ and ΓX directions for an FCC periodic composite consisting of solid spheres in water host. The volume fractions of the spheres are $f_s = 50\%$. Figure 14.1b shows the band structure for fluid spheres of the same λ and ρ as the solid ones and in the same periodic arrangement also in water host. Figure 14.1b shows the band structure results for the same material combination using the plane wave (PW) method. The results are very different, which shows the replacement of solid scatterers with fluids can change the band structure drastically.

The results for a different material system consisting of glass spheres are shown in Fig. 14.2. This is in SC structure and glass volume fraction $f = 45\%$. Again here, the left panel is MS method result and the right is using PW method. Compared with Fig. 14.1, the difference between the MS result and the PW result is reduced. This means that reduction of the influence of the rigidity of the scatterers can be attributed to the larger velocity and density contrast between scatterers and host. The velocity and mainly the density contrast between scatterers and host are the

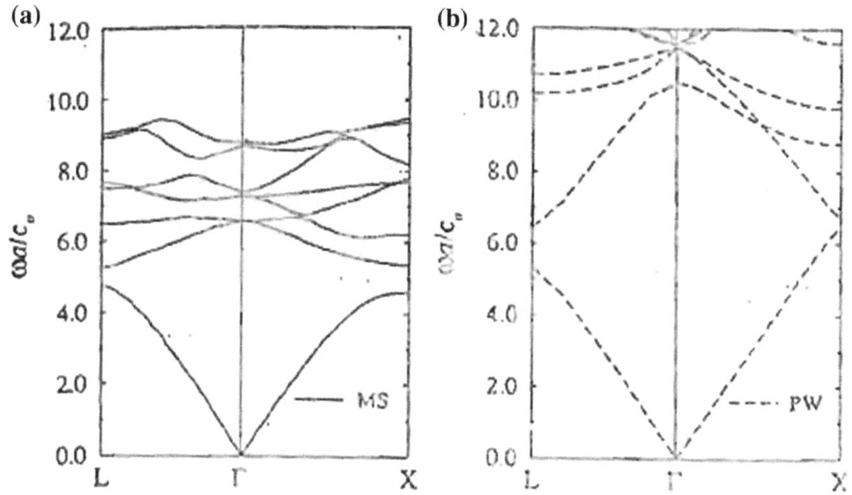

Fig. 14.1 Dispersion relation along the LΓ and ΓX directions for an FCC periodic composite consisting of solid spheres in water. The parameters are as follows: $\frac{\rho_o}{\rho_n} = 1/2$, $\frac{c_o}{c_{\ell i}} = 1/2$, $\frac{\lambda_o}{\lambda_i} = 1/4$, $\frac{c_{ti}}{c_{\ell i}} = 1/2$. Volume fractions of spheres $f = 50\%$. c_o is the wave velocity in the host and a the lattice constant, **a** shows the result within MS method and **b** the same with PW and Eq. (15.3.1) (From Kafesaki and Economou [18])

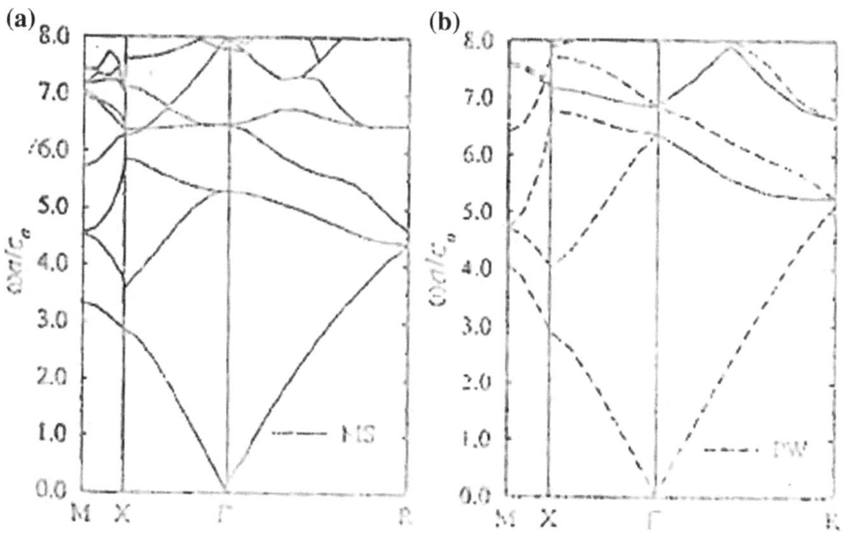

Fig. 14.2 Dispersion relation along the MXΓR directions for a SC periodic composite consisting of glass spheres in water. Glass volume fraction $f = 45\%$. c_o is the wave velocity in the water and a the lattice constant (From Kafesaki and Economou [18])

most important parameters controlling the scattering and thus the propagation in the composite system. As these contrasts increase, other parameters as the rigidity of the scatterers become less important [23, 24].

14.5 Negative Refraction Derived from Gauge Invariance (Coordinate Transformations)—An Alternative Theory of Negative Refraction

14.5.1 Gauge Invariance as a Unified Theory of Negative Refraction and Cloaking

Phononic crystals as a slab of periodically arranged scatterers can produce perfect lens beating the Rayleigh diffraction limit and produce negative refraction effect [25]. Slab of phononic crystals combined with ordinary material will produce new reflection coefficient and transmission coefficient due to their capabilities of sound reduction and vibration reduction.

Here, we consider the refraction of a sound wave at the boundary between two media with different parities. That is, one is a double-positive (DPS) material with positive mass density and positive bulk modulus and hence the parity is $+1$; the other medium is a double-negative (DNG) material with negative mass density and negative bulk modulus, and hence the parity is -1. The acoustic equation of motion is invariant in the form (gauge invariant) to both parity equals $+1$ and -1. We call this parity invariance.

The x and y components of the fields $\vec{V}$ and $\vec{T}$ in the refracted ray maintain their directions, independently of the parities of the two media ($\vec{V}$ = particle velocity and $\vec{T}$ = stress field). As for the z component, it keeps the same direction only if the two media are of the same parity. If the parities are different, the z components change sign. This corresponds to the fact that in passing into a medium of different parities, the vectors $\vec{V}$ and $\vec{T}$ not only change in magnitude according to the difference in ρ and κ (ρ = mass density and κ = bulk modulus or compressibility) but also undergo a reflection relative to the interface of the two media. The same thing happens to the wave vector $\vec{k}$ too. The simultaneous reflection of all three vectors corresponds precisely to a change of sign of the determinant of

$$G = \begin{pmatrix} \alpha_1 & \alpha_2 & \alpha_3 \\ \beta_1 & \beta_2 & \beta_3 \\ r_1 & r_2 & r_3 \end{pmatrix} \tag{14.14}$$

The determinant of matrix G is the parity p. $\alpha_i, \beta_i, r_i (i = 1, 2, 3)$ = direction cosines for the vectors $\vec{\rho}$, $\vec{\kappa}$, and $\vec{k}$ and G characterize the wave propagation in the medium [26].

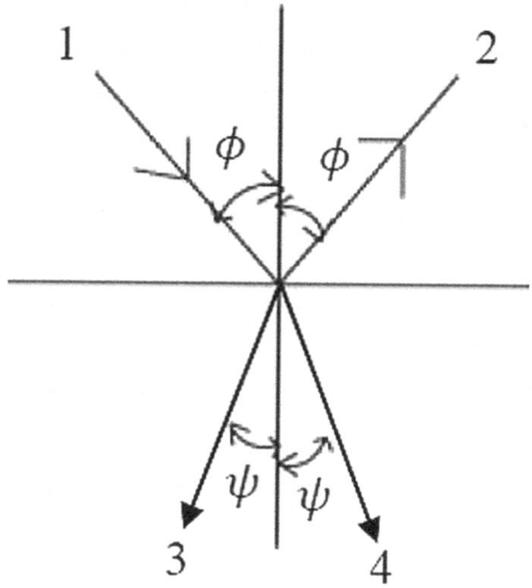

Fig. 14.3 Passage of a sound wave through the boundary between two media. 1—incident ray; 2—reflected rays; and 3—refracted rays if the second medium is left-handed; 4—refracted rays if the second medium is right-handed

The path of the refracted wave produced as the result of such reflections is shown in Fig. 14.3 [26].

There will be negative refraction when there is double negativity and the parity equals -1.

Negative refraction can also be obtained from coordinate transformations (gauge invariance). Coordinate transformations in space consist of either a translation or a rotation or both. In applying coordinate transformations, the underlying principle [27] states that it is well known that coordinates of an arbitrary point or a vector in the new system (x', y', z') can be specified in terms of the old system (x, y, z) as

$$\begin{bmatrix} x' \\ y' \\ z' \end{bmatrix} = \begin{bmatrix} \alpha_1 & \alpha_2 & \alpha_3 \\ \beta_1 & \beta_2 & \beta_3 \\ r_1 & r_2 & r_3 \end{bmatrix} \begin{bmatrix} x \\ y \\ z \end{bmatrix} \tag{14.15}$$

The negative refraction is a special case of coordinate transformations when the determinant of the direction cosines matrix (transformation matrix) in (14.15) equals -1. This also produces the concept of negative mass density and negative compressibility for negative refraction as can be shown as follows:

$$\begin{aligned} [\rho'_{i\ell}] &= (-1)[\rho_{i\ell}] \\ [\kappa'_{i\ell}] &= (-1)[\kappa_{i\ell}] \end{aligned} \tag{14.16}$$

Hence, gauge invariance or coordinate transformations is a unified theory of negative refraction and cloaking.

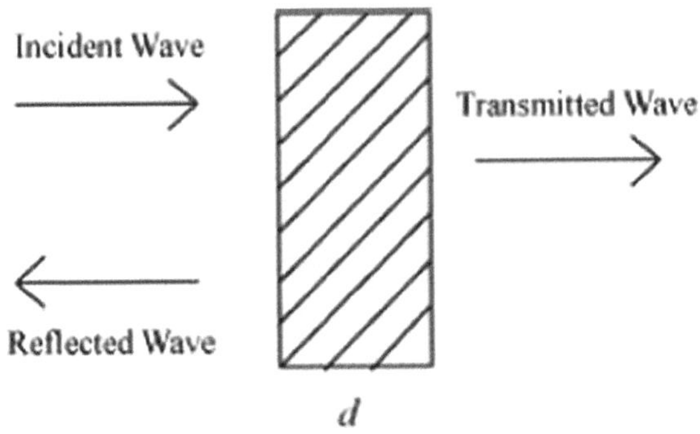

Fig. 14.4 Plane wave scattering from a DNG slab of thickness d

Another illustration of the unified theory is that when we replace the mass density and the bulk modulus by the forms altered by coordinate transformations, there is no change in the form of the acoustic equation of motion. Likewise, if we replace the mass density and the bulk modulus by the negative mass density and negative bulk modulus, again there is no change in the form of the acoustic equation of motion. This shows that negative refraction is a special case of the coordinate transformations.

From Fig. 14.3, we realize that when the second medium is left-handed or parity equals -1, the refracted ray lies on the opposite side of the z-axis from its position in the case of a right-handed second medium [28]. It must be noted that the direction of the reflected sound wave is always the same, independent of the parities of the two media. Figure 14.3 shows that Snell's law of refraction for acoustics has to be modified by the introduction of the parity p:

$$\frac{\sin\phi}{\sin\psi} = n_{1,2} = \frac{p_2}{p_1}\sqrt{\frac{\rho_2\kappa_1}{\rho_1\kappa_2}} \tag{14.17}$$

where $\rho_2, \kappa_2 =$ mass density and bulk modulus of the second medium; $\rho_1, \kappa_1 =$ mass density and bulk modulus of the first medium; $p_2 =$ parity of the second medium; and $p_1 =$ parity of the first medium. The index of refraction of two media can be negative if the parities of the media are different. In particular, the index of refraction of a left-handed medium relative to vacuum is negative [1].

The concept of negative refraction has been utilized in the fabrication of superlens which defeats Rayleigh's resolution limit [29] which has the capability of gradually converting evanescent components into propagating waves.

Here, we introduce Pendry's [30] idea for subwavelength focusing and perfect lensing into acoustics. We consider the scattering from a DNG slab. The reflection and transmission coefficients associated with a normally incident plane wave that scatters from a DNG slab embedded in a medium have been derived as follows:

The DNG slab has an infinite extent in the transverse directions, with a thickness d in the direction of propagation of the incident plane wave (Fig. 14.4).

14.5.2 Generalized Form of Snell's Law for Curvilinear Coordinates

So far, Snell's law has been derived for Cartesian coordinates. In order for the manipulation and control of the sound wave propagation, it would be necessary to derive Snell's law for curvilinear coordinates using coordinate transformations (gauge invariance). In (14.5.4), the parity p will be replaced by the determinant of the direction cosines matrix. A practical example for the application of Snell's law in curvilinear coordinates can be illustrated as follows as quoted from Stephen Hawking and Leonard Mlodinow's book The Grand Design [31].

A few years ago, the city council of Monza, Italy barred pet owners from keeping goldfish in curved goldfish bowls. The sponsor explained the measure in part by saying that it is cruel to keep a fish in a bowl with curved sides because gazing out, the fish would have a distorted view of reality. Due to the distortion, a freely moving object that we would observe to move in a straight line would be observed by the goldfish to move along a curved path. Nevertheless, the goldfish could formulate scientific laws from their distorted frame of reference that would always hold true and that would enable them to make predictions about the future motion of objects outside the bowl. Their laws would be more complicated than the laws in our frame but simplicity is a matter of taste.

14.5.3 Design of a Perfect Lens Using Coordinate Transformations

Wang et al. [32] designed oblate cylindrical perfect lens using coordinate transformations instead of using negative refraction approach. This further confirms our unified theory of negative refraction and coordinate transformations.

14.5.4 A General Cloaking Lens

We will start with electromagnetic wave. Here, we extend the concept of negative refraction and perfect lens to a general cloaking lens based on the forms of permittivity and permeability produced by our predetermined direction of wave propagation using coordinate transformations. This will go beyond controlling and manipulating refraction to controlling and manipulating diffraction. An ordinary optical lens is made of optical material which is able only to control refraction because only interaction of electric field with optical material is possible. However, with the fabrication of metamaterial with negative permeability, it allows the interaction of the magnetic field with the metamaterial and hence the control and the manipulation of diffraction or the bending of the light wave becomes possible. This allows the fabrication of a cloaking lens which is now controlling light wave with a nonlinear path based on the curvilinear coordinates instead of the optical lens which is controlling only light waves of straight path based on the Cartesian coordinates. The shapes of the cloaking lens can be from a rectangular slab of metamaterial to spherical and cylindrical shapes.

Likewise, we can extend the same concept to the sound wave. Here, we can fabricate a general cloaking lens made of general forms of mass density and bulk modulus decided by our predetermined direction of the sound wave propagation beyond the perfect lens and negative refraction which are based only on the special case of double-negative material.

The concept of a general cloaking lens is another illustration of the idea that negative refraction is a special care of coordinate transformations.

14.6 Reflection and Transmission of the Sound Wave at Interface of Two Media with Different Parities

Let the medium before and after the slab be characterized by ρ_1, κ_1 and ρ_2, κ_2, respectively. For a normally incident plane wave, the reflection and transmission coefficients for the slab are:

$$R = \frac{z_2 - z_1}{z_2 + z_1} \frac{1 - e^{-j2k_2d}}{[(z_2 - z_1)/(z_2 + z_1)]^2 e^{-j2k_2d}} \tag{14.18}$$

$$T = \frac{z_2 z_1}{(z_2 + z_1)^2} \frac{e^{-j2k_2d}}{1 - [(z_2 - z_1)/(z_2 + z_1)]^2 e^{-j2k_2d}} \tag{14.19}$$

where z = acoustic impedance = $\sqrt{\rho c_{44}}$ and K = wave vector = $\omega\sqrt{\frac{\rho}{\kappa}}$.

For the case of normal incidence, and if we consider a matched DNG medium, then $z_2 = z_1$ and $R = 0$, and $T = e^{-jk_2d} = e^{+j|k_2|d}$. The medium would add a positive phase to the wave traversing the slab, whereas in a DPS medium the wave would experience a negative phase variation. This means that a matched DNG could be used to compensate for phase changes incurred by passage of a plane wave through a DPS slab. That is, one can have $k_{\text{DPS}}d_{\text{DPS}} = k_{\text{DNG}}d_{\text{DNG}} = 0$. This phase compensation is an interesting feature of DNG metamaterials that can lead to exciting potential applications.

With an obliquely incident plane wave, Eqs. (14.18) and (14.19) can be modified by introducing the transverse impedance and longitudinal wave components. If, in addition, the incident wave is evanescent, that is, when the transverse component of a wave vector of the incident wave is greater than the wave number of the medium ($k_t^2 > \omega^2 \frac{\rho_1}{k_1}$ and $k_t^2 > \omega^2 \frac{\rho_2}{k_2}$), the transverse wave impedance in each medium (with a no-loss assumption) becomes purely imaginary, that is $Z_{1,\text{transverse}} = jX_{1,\text{transverse}}$ and $Z_{2,\text{transverse}} = jX_{1,\text{transverse}}$, and the longitudinal component of the wave vector in each medium also becomes purely imaginary, that is, k_1, longitudinal $= j\alpha_1$, and k_2, longitudinal $= j\alpha_2$. However, it can be shown that this transverse wave impedance in the DPS and DNG media has opposite signs. That is, if one has a capacitive reactance, the other will have an inductive reactance so that $\mathrm{sgn}\left[X_{1,\mathrm{transverse}}\right] = -\mathrm{sgn}\left[X_{2,\mathrm{transverse}}\right]$, where $\mathrm{sgn}(x) = +1(-1)$ for $x > 0$ ($x < 0$). When we choose the matched condition for which $\rho_2 = -\rho_1$ and $k_2 = -k_1$, one can demonstrate that $X_{1,\mathrm{transverse}} = -X_{2,\mathrm{transverse}}$. Inserting these features into the generalized form of Eqs. (14.18) and (14.19), one would obverse that again $R = 0$ but now $T = \mathrm{e}^{-jk_{(2,\mathrm{longitudinal})}d} = \mathrm{e}^{\alpha_2 d}$. It is known that at the interface between a DPS and a DNG medium the tangential components of the particle velocity field and the stress fields should be continuous according to the acoustic field equation. However, since the mass density and the bulk modulus of these two media both have opposite signs, the normal spatial derivative (normal with respect to the interface) of these tangential components is discontinuous at this boundary. Moreover, according to Eqs. (14.18) and (14.19), the overall reflection coefficient for the incident evanescent wave reaches the first interface of the matched DNG slab from the DPS region, it is decaying, that is $\alpha_1 < 0$, and no reflected evanescent wave will be present in this DPS medium. However, as we moved just past the first interface into the DNG region, the tangential components of the field in the vicinity of the interface inside the DNG region should grow in order to satisfy the discontinuity condition of the normal spatial derivative mentioned above. Note that if the evanescent wave decayed inside this matched DNG slab, the tangential components of the field at the DPS–DNG interface would have similar slopes, inconsistent with the boundary condition mentioned above. Therefore, in the transmission coefficient expression, $T = \mathrm{e}^{\alpha_2 d}$, one should have $\alpha_2 > 0$. As a result, such a matcher DNG slab can compensate for the decay of the evanescent wave in the DPS region through the growth of the evanescent wave inside the DNG slab. This is the basis behind the idea of subwavelength focusing and perfect lensing.

This scenario represents the presence of an interface resonance at the boundary between the DPS and the DNG regions. Furthermore, this phenomenon can also be described and justified using distributed circuit elements.

14.7 Theory of Diffraction by Negative Inclusion

14.7.1 Formulation of Forward Problem of Diffraction Tomography

The diffraction in a double-negative (DNG) medium can be studied following the formulation of diffraction tomography. The ability to control and manipulate diffraction will give rise to a new form of diffraction and a new theory of diffraction. We will consider the diffraction by negative inclusion which described a double-negative medium compared with the conventional diffraction tomography which deals with a positive medium with parity equal to +1. We shall follow the approach of Burov et al. [33] who used the linearized hydrodynamic equation as the starting point instead of the Helmholtz wave equation to derive the Lippmann–Schwinger equation used in diffraction tomography. The reason is the Helmholtz wave equation involves the refractive index squared. This is not quite consistent and requires additional justifications concerning the sign of n in a left-handed medium that is associated with the necessity of doing a positive work by the source, with the direction of the vector of the energy flux away from the source, with the causality principle, etc. We have to resort to the causality principle to justify the negative sign of the refractive index for a negative medium [34]. The linearized hydrodynamic equation on the other hand does not involve inconsistency arises in choosing the sign of n, the refractive index.

The linearized hydrodynamic equation (for acoustics) is:

$$\begin{aligned} &\frac{\delta}{\delta t}\left(\widehat{K}p\right) + \nabla\vec{V} = \varphi \\ &\frac{\delta}{\delta t}(\hat{\rho}p) + \nabla\vec{p} = \vec{f} \end{aligned} \tag{14.20}$$

where φ and $\vec{f}$ are the scalar and vector primary sources of the acoustic field, respectively. For media without dispersion, $\vec{p}$ and $\widehat{K}$ are scalars. When dispersion is present, they are convolution-type operators over the time variable.

A four-dimensional representation of the field variables (pressure p and three components of velocity $\vec{V}$) is used in (14.20). The resonator response function $Q(\tau)$ which characterizes the appearance of a secondary source in response to the incident field is then a 4×4 matrix at each fixed τ. The properly transformed Kramers–Kronig relations also acquire a matrix form.

Since ρ and κ can be negative only in a narrow frequency band, whose width is determined by the resonator Q factor, Burov et al. [33] discuss only stationary monochromatic fields at this stage. As a result, they can abstract from the specific resonator design and disregard the explicit form of the response function $Q(\tau)$, reasoning in the language of effective functional parameters ρ and κ, which depend only on the coordinates at a fixed frequency. It should be noted that the monochromatic case cannot give an answer to the question about the relationship between the real and the imaginary parts of each functional parameter characterizing the medium since this relationship follows precisely from the Kramers–Kronig relations. However, this relationship should also be taken into account when working with monochromatic fields.

An ordinary (positive) medium is considered as a background with parameters ρ_o and κ_o, while a negative medium is obtained from it by adding corrections $\rho'(\vec{r})$ and $\kappa'(\vec{r})$ (which are not small): $\rho(\vec{r}) \equiv \rho_0 + \rho'(\vec{r})$, $\kappa(\vec{r}) \equiv \kappa_o + \kappa'(\vec{r})$. This approach allows the propagation of waves in media with arbitrary (in magnitude and sign) ρ and κ distributions to be calculated using the well-known methods of the scattering theory that are not based on the Born approximation or similar assumptions.

In the monochromatic case, the system of equations (14.20) for time dependence $\exp(-i\omega t)$ transforms to

$$\nabla\vec{V} - i\omega q p = \varphi;\quad \nabla p - i\omega\rho\vec{V} = \vec{f} \tag{14.21}$$

Burov et al. [32] introduce column vectors $\check{u} \equiv \begin{bmatrix} \vec{V} \\ p \end{bmatrix}$, $\check{F} \equiv \begin{bmatrix} \vec{f} \\ \varphi \end{bmatrix}$ and an operator $\widehat{A}$ in the form of a matrix acting both in the coordinate space and in the space of field variables $\begin{bmatrix} \vec{V} \\ p \end{bmatrix}$:

$$\widehat{A} \equiv \begin{bmatrix} -i\omega\rho(\vec{r}) & \nabla \\ \nabla & i\omega K(\vec{r}) \end{bmatrix} = \widehat{A}_o - \widehat{A},$$

where $\widehat{A}_o \equiv \begin{bmatrix} -i\omega\rho_o & \nabla \\ \nabla & i\omega K_o \end{bmatrix}$ and $\widehat{A}_1 \equiv \begin{bmatrix} i\omega\rho'(\vec{r}) & 0 \\ 0 & i\omega K'(\vec{r}) \end{bmatrix}$ are the operators that characterize the homogeneous medium of a positive background and its perturbation, respectively. It should be noted that the introduced quantities are a combination of both scalar and vector field components in the region under study. In matrix form, system (14.21) is $\widehat{A}\vec{u} = \check{F}$. At $\widehat{A}_1 = 0$, implying the absence of corrections $\rho'(\vec{r})$ and $K_o'(\vec{r})$, the sources $\check{F}$ produce the incident field $\check{u}_o \equiv \begin{bmatrix} V_o \\ P_o \end{bmatrix}$, in the region under consideration, and the system is

$$\widehat{A}_0\check{u}_o = \check{F} \tag{14.22}$$

hence

$$\vec{u}_o = \widehat{A}_o^{-1}\vec{f} \tag{14.23}$$

Let $\widehat{G}$ be the retarded Green function of a homogeneous medium for the system of (14.22), i.e. a system of type (14.21),

$$\widehat{A}_o^{-1}(\cdot) = \int \widehat{G}(\vec{r} - \vec{r}')(\cdot)\mathrm{d}\vec{r}'$$

In the presence of corrections ρ' and κ', the following relation holds for the $\check{u}$:

$$\vec{u} = \widehat{A}^{-1}\vec{f} = \widehat{A}^{-1}\widehat{A}_0\widehat{A}_o^{-1}\vec{f} = \left[\widehat{A}^{-1}\widehat{A}_0\right]\vec{u}_o$$
$$= \left[\widehat{A}^{-1}\widehat{A}\right]^{-1}\vec{u}_o = \left[\widehat{A}_o^{-1}\left(\widehat{A}_0 - \widehat{A}_1\right)\right]^{-1}\vec{u}_o = \left[\widehat{E} - \widehat{A}_o^{-1}\widehat{A}_1\right]^{-1}\vec{u}_o$$

Hence $\vec{u} = \left[\widehat{E} - \widehat{G} * \widehat{A}_1\right]^{-1}\vec{u}_o$ (14.24) where $\widehat{E}$ = unit operator and $*$ denotes the convolution operation in the coordinate space.

Equation (14.24) is an operator form of the solution to the Lippmann–Schwinger equation for the field $\vec{u}$:

$$\vec{u}(\vec{r}) = \vec{u}_o(\vec{r}) + \int\limits_R \widehat{G}(\vec{r} - \vec{r}')\left[\widehat{A}_1(\vec{r}')\vec{u}(\vec{r}')\right]\mathrm{d}\vec{r}' \tag{14.25}$$

where R is the localization region of the ρ' and κ' inhomogeneities.

Here, the operator $\widehat{A}_1(\vec{r}')$ acts in the space of wave variables $[\vec{p}]$, while in the coordinate space it is a local multiplication operator at each point $\vec{r}$. The inverse operator in (14.24) exists since all its eigenvalues are complex for passive media, to which the media under consideration belong [35] and impose no constraints on the scatterer strength.

An explicit expression for the matrix form of the retarded Green function $\widehat{G}$ can be derived [36] by passing the space of wave vector $\vec{K}$, i.e. by decomposing the fields using a set of plane harmonic waves for which the pressure and oscillation velocity vary as $\exp\left(-i\omega t + i\vec{K}\vec{r}\right)$. In this space, the operator $\widehat{A}_o$ takes the form of an ordinary matrix $\begin{bmatrix} -i\omega\rho_o & i\vec{K} \\ i\vec{K} & -i\omega\kappa_o \end{bmatrix}$, whose inversion gives $\begin{bmatrix} -i\omega\rho_o & i\vec{K} \\ i\vec{K} & -i\omega\kappa_o \end{bmatrix}^{-1}$ $= \frac{1}{K_o^2 - \vec{K}^2}\begin{bmatrix} -i\omega\kappa_o & i\vec{K} \\ i\vec{K} & -i\omega\rho_o \end{bmatrix}$, where $k_o = \omega\sqrt{\rho_o\kappa_o}$ is the wave number in the background medium. To obtain the Green function $\vec{G}$ in coordinate representation, we must perform the inverse Fourier transform (the transition from $\vec{K}$ to $\vec{r}$) of this expression. In this case, a pole appears in the denominator going around which requires introducing an infinitesimal imaginary part $\pm\xi$ for the wave number: $k_o = \omega\sqrt{\rho_o\kappa_o} \pm i\xi$, where $\xi \to +0$. The "+" or "−" sign in front of the infinitesimal but positive ξ defines whether the Green function will be retarded or advanced one. As a result, for the retarded Green function, we derive the expression $\widehat{G}(\vec{r} - \vec{r}')$ $\equiv \begin{bmatrix} -i\omega\rho_o & \nabla \\ \nabla & i\omega K_o \end{bmatrix} G(\vec{r} - \vec{r}')$, where $G(\vec{r} - \vec{r}')$ is the retarded Green function of the Helmholtz equation for a homogeneous space of the corresponding dimension with parameters ρ_o and κ_o, whose analytical form is well known. The operator $\nabla \equiv \nabla_r$ acts on the argument $\vec{r}$ of the function $G(\vec{r} - \vec{r}')$: in the one-dimensional ($D = 1$) and two-dimensional ($D = 2$) cases, the Green function for the system of equations (14.21) is then

$$\widehat{G}_{D=1}(x - x') = \frac{\exp(ik_o|x - x'|)}{2}\begin{bmatrix} \sqrt{\frac{k_o}{\rho_o}} & \mathrm{sgn}(x - x') \\ \mathrm{sgn}(x - x') & \sqrt{\frac{\rho_o}{k_o}} \end{bmatrix} \tag{14.26}$$

$$\widehat{G}_{D=2}(\vec{r}-\vec{r}\,')=\frac{i}{4}\begin{bmatrix} -i\omega\kappa_0 H_0^{(1)}\left(K_0\left|\vec{r}-\vec{r}\,'\right|\right) & \frac{\vec{r}-\vec{r}\,'}{\left|\vec{r}-\vec{r}\,'\right|}\kappa_0 H_1^{(1)}\left(K_0\left|\vec{r}-\vec{r}\,'\right|\right) \\ \frac{\vec{r}-\vec{r}\,'}{\left|\vec{r}-\vec{r}\,'\right|}\kappa_0 H_1^{(1)}\left(K_0\left|\vec{r}-\vec{r}\,'\right|\right) & -i\omega\rho_0 H_0^{(1)}\left(\kappa_0\left|\vec{r}-\vec{r}\,'\right|\right) \end{bmatrix} \quad (14.27)$$

Equations (14.24) and (14.25) are written for the system of hydrodynamic equations (14.21) and not for the Helmholtz wave equation.

Since the background medium in Eqs. (14.24) and (14.25) is positive, there is no need to use the advanced Green function to calculate the field in a negative medium. These relations allow the field to be reproduced for any configuration of a finite (in size) scatterer that consists of a positive or negative material for an arbitrary incident field $\check{u}_0$.

One has to discretize Eq. (14.25). The region under consideration is sampled by dividing it into areas δS_n, characterized by the radius vectors of their centres $\vec{r}_n$. The area size is chosen to be much smaller than the wavelength, in such a way that the parameters of the medium ρ and κ, within each area as well as the incident field $\check{u}_0$ and the diffracted field $\check{u}$, may be considered constant. The integration on the right-hand side of Eq. (14.7.6) is reduced to the summation over the area δS_n, in each of which $\widehat{A}_1$ and $\check{u}$ are assumed to be constants determined only by the area number n. For the field inside area m, the sampled form of the Lippmann–Schwinger equation becomes:

$$\begin{aligned} \vec{u}_m &= \vec{u}_{om} + \sum\left[\int\limits_{\delta S_n} \widehat{G}(\vec{r}_m-\vec{r}\,')\widehat{A}_1(\vec{r}\,')\vec{u}(\vec{r}\,')\mathrm{d}\vec{r}\,'\right] \\ &\approx \vec{u}_{om} + \sum_n\left[\int\limits_{\delta S_n} \widehat{G}(\vec{r}_m-\vec{r}\,')\mathrm{d}\vec{r}\,'\right]\left[\widehat{A}_1\vec{u}\right]_n \end{aligned} \quad (14.28)$$

From Eqs. (14.26) and (14.27), the Green function is smooth everywhere in the one-dimensional case and at $\vec{r}\neq\vec{r}\,'$ in the two-dimensional case. The amount of calculations can then be reduced significantly when calculating the elements of the matrix $\widehat{G}_{mn}\approx\int_{\delta S_n}\widehat{G}(\vec{r}_m-\vec{r}\,')\mathrm{d}\vec{r}\,'$, by assuming it to be constant within the area and equal to its value at the area's centre in these cases:

$$\widehat{G}_{mn}\approx\widehat{G}(\vec{r}_m-\vec{r}_n)\delta S_n \quad (14.29)$$

Equation (14.29) cannot be used to calculate the matrix elements $\widehat{G}_{mn}$ in the two-dimensional and three-dimensional case at $m=n$ and must perform the integration over the area $\delta S_n=m$. In this case, the singularity of the function $\widehat{G}$ that arises when its argument tends to zero (in the two-dimensional case, the Hankel functions $H_0^{(1)}$ and $H_1^{(1)}$ have a singularity) is integrable.

As a result of the estimation of all matrix elements $\widehat{G}_{mn}$, Eq. (14.7.9) takes the form

$$\vec{u}_m = \vec{u}_{om} + \sum_n \widehat{G}_{mn}\left[\widehat{A}_1 u\right]_n \tag{14.30}$$

The following is the sampled analogue of Eq. (14.24):

$$\vec{u}_m = \left[E_{nm} - \left[\widehat{G}\widehat{A}_1\right]_{nm}\right]^{-1} \vec{u}_{on} \tag{14.31}$$

In Eq. (14.31), the matrix elements $\left[\widehat{G}\widehat{A}_1\right]_{nm}$ are formed by the products of the matrices $\widehat{G}_{mn}$ and $\left[\widehat{A}_1\right]_m$ at fixed m. All quantities in Eqs. (14.24) and (14.25) and (14.28) to (14.31) are defined on the direct product of the space of field variables and the coordinate space of vector $\vec{r}$ sampled in terms of the indices m and n. Hence, it is appropriate to use Eq. (14.31) to find the field inside the scatterer localization region R and in its neighbourhood with minimum sizes required by the problem, since in the case of an unjustifiable increase in the sizes of the region being analysed, the matrix inversion operation requires a very large amount of a computer's random access memory.

It has to be noted that during the inversion of the nonsampled operator in Eq. (14.24), the width of its spatial spectrum increases, since the inversion procedure reflects the entire set of rescattering process [34]. Therefore, even at the formation stage of the direct matrix $E_{nm} - \left[\widehat{G}\widehat{A}_1\right]_{nm}$, one has to significantly reduce the spatial sampling step for the inversion of the matrix in Eq. (14.31) to lead to the correct values of $\breve{u}_m$. The optimal choice of the sampling step will be considered during the process of numerical simulation. Once the internal field $\breve{u}_n$ and consequently the secondary sources of the scattered field have been found at all points of region R, the total field $\breve{u}_m$ out of any point $\vec{r}_m \notin R$ can be found as the sum of the incident field from the primary sources and the scattered field from the sampled secondary sources $\left[\widehat{A}_1\breve{u}\right]_n$ from Eq. (14.25), which after the sampling takes a form similar to Eq. (14.28):

$$\breve{u}_m^{\text{out}} = \breve{u}_{om}^{\text{out}} + \sum_n \widehat{G}_{mn}\left[\widehat{A}_1\breve{u}\right]_n \tag{14.32}$$

Hence, Burov et al. [33] have constructed a mathematical framework suitable for modelling the wave processes in both positive and negative media.

14.7.2 Modelling Diffraction Procedure in a Negative Medium

In diffraction to tomography, one is concerned with the propagation of the sound wave in an inhomogeneous medium which in this case is the composite double-negative (DNG) material. This will involve refraction, reflection, transmission and multiple scattering processes at the boundaries. Bliokh et al. [37] argued that a two-dimensional medium with negative refraction of a wave at a boundary can be considered as a left-handed medium. However, there exist several more effects that allow one to talk about a negative medium even in the one-dimensional case. First, these include the wave reflection and transmission coefficients in pressure and oscillation velocity in combination with information about the sign of the effective density of the medium. Thus, for example, in the special case of a perfect match between the bordering positive (with density ρ_0 and phase velocity c_0) and the negative (ρ and c) acoustic media, when their density and compressibility relations $\rho = \rho_0$ and $\kappa = -\kappa_0$ hold, the impedances $\rho c = \rho_0 c_0$ and the absolute values of the phase velocities of these media coincide. Owing to the coincidence of the impedances, there is no wave reflected from the boundary. Thus, while talking about a one-dimensional negative medium, the phase velocity c in it may be considered from the absence of a reflected wave and a priori knowledge about the negativity of the effective density $\rho < 0$ to be also negative. Secondly, the conclusion about whether the directions of the energy propagation and the phase velocity are the same or opposite can be drawn directly from the following effect.

For a plane monochromatic acoustic wave with a wave vector $\vec{k}_0$, propagating along the Z-axis (pressure and velocity $P_0, v_0 \sim \mathrm{e}^{i\vec{k}_0\vec{r}} = \mathrm{e}^{\pm ik_0 z}$), the three-dimensional plot with the imaginary and real parts of the pressure or oscillation velocity along the X- and Y-axes, respectively, and with the only spatial coordinate along the z-axis is a left-handed or right-handed helix, depending on whether the wave vector and hence the phase velocity are oriented in the negative (Fig. 14.5; $P_0 v_0 \sim \mathrm{e}^{-ik_0 z}$) or positive direction of the Z-axis. In other words, for a fixed time dependence $\sim \mathrm{e}^{-i\omega t}$, there is a unique relationship between the orientation of vector $\vec{k}_0$ and the sign of helicity; i.e. the sign of HELIV) is determined by the direction of $\vec{k}_0$ in a given medium, positive or negative. If two waves propagate in the opposite directions in a positive or a negative medium, then the circular helix becomes an elliptical one or even concentrates in the form of oscillations on a plane if the amplitudes of both waves are equal. Since the direction of rotation of the circular helix is uniquely related to the direction of the phase velocity in the medium, the change of the sign of helicity at the boundary between two media is a criterion that one of them is negative.

Since Eqs. (14.31) and (14.32) suggest the field construction simultaneously in the entire scattering region R, scatterers with finite sizes are considered in numerical simulations. A background medium with $\rho_0 \equiv 1$ and $\kappa_0 \equiv 1$ is assumed to be unbounded, and the incident field $\breve{u}_0$ is defined analytically everywhere. In the numerical simulations of Burov et al. [38], they considered a layer of the background medium with a thickness of $5\lambda_0$ where λ_0 is the wavelength in the background medium. In turn, a thinner layer of a positive ($\rho \equiv 2$ and $\kappa \equiv 5$) or perfectly matched (with the background medium) negative ($\rho \equiv -1$ and $\kappa \equiv -1$) material with a thickness of $2\lambda_0/3$ was placed in the middle part of this layer. The sampling step was $\lambda_0/100$. The acoustic pressure field of the incident plane wave P_0 which is described by one

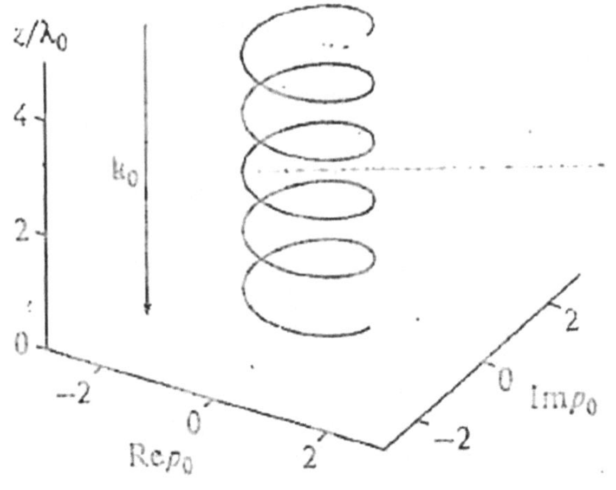

Fig. 14.5 Graphic representation of the incident monochromatic wave. The arrow indicates the direction of the wave vector $\vec{k}_0$ (From Barov et al. [33])

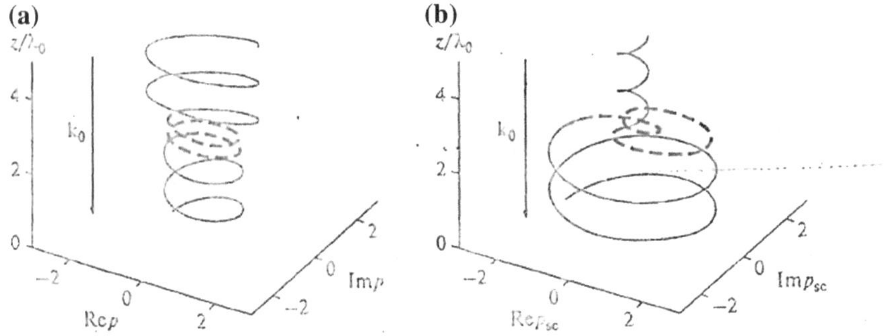

Fig. 14.6 Calculated total acoustic pressure field p (**a**) and scattered field p_{SC} (**b**) in a layer of a positive material. The solid and dashed lines indicate the fields in the background positive medium and in the layer placed in it, respectively. The arrow indicates the direction of the wave vector $\vec{k}_0$ of the incident wave (From Barov et al. [33])

component in the one-dimensional case of normal incidence on the layer here is shown in Fig. 14.5.

14.7.3 Results of Numerical Simulation

The total pressure p and the scattered field $p_{\mathrm{SC}} \equiv p - p_0$ calculated when a layer of a positive material was included in the background medium are presented in Fig. 14.6.

The helix corresponding to the total pressure field p (Fig. 14.6a) changes its shape (but not the sign of helicity) at the boundaries of the layer, because the waves reflected from the boundary appear. Figure 14.6b shows the field scattered by the layer. It shows that the helix corresponding to the wave reflected from the layer has an inverse helicity with respect to the incident field (Fig. 14.5), since the wave vectors of the incident and reflected waves are directed oppositely. The radius of this helix, which is equal to the amplitude of the reflected wave $|R_p p_0|$ (here $|p_0| = 1$), shows the magnitude of the reflection coefficient from the layer in pressure $|R_p|$ to be estimated. This coefficient can be compared with its theoretical value calculated from the formula [36, p. 40]

$$|R_p| = \frac{|s^{-1} - s|}{\sqrt{(s^{-1} - s)^2 + 4\cot^2(kd)}} \tag{14.33}$$

where $S \equiv \frac{\rho_0 c_0}{\rho c}$, k and c are the wave number and the speed of sound in the medium, k_0 and c_0 are those in the background medium, and d is the layer thickness. For $\rho \equiv 2$ and $\kappa \equiv 5$, $|R_p|$ can be analytically obtained as approximately 0.3588. The magnitude of the reflection coefficient is calculated using the data from Fig. 14.6b $|R_p| = 0.3576$, corresponding to its theoretical value within the limits of the errors caused by sampling. This is indicative of an adequate qualitative and quantitative description of the fields with Eqs. (14.24) and (14.25).

Figure 14.7 presents similar plots for an inclusion in the form of a layer of a negative material. Note that the sign of helicity of the total field changes (Fig. 14.7a) at the boundaries of the layer. This means that the phase velocity of sound in a negative medium is opposite in direction to that in positive one. It should also be noted that the radius of the helix of the scattered field (Fig. 14.7b) on the segment of the Z-axis from $2.7\lambda_0$ to $5\lambda_0$ is 0 and, hence, there is no reflection from the layer at $\rho \equiv 1$ and $\kappa \equiv 1$, as expected, owing to a perfect match between the background medium and the medium of a negative material with identical acoustic impedances. Since there is no reflected wave, the Poynting vector $\vec{S}$, along with the group velocity, is constant in the entire segment under consideration, and are directed along the Z axis. Thus, the wave vector $\vec{k}$ is directed along $\vec{S}$ in a positive medium and opposite to it in a negative one. As a result, as the wave energy propagates along the vector $\vec{S}$, the coordinate phase shift of the wave has opposite signs in positive and negative media.

The numerical simulations above for the one-dimensional case show that the simultaneous negativity of the density and compressibility of the material is a sufficient condition that the medium has a number of properties usually attributed to negative refraction.

The next step is to consider two-dimensional model of a negative medium, which allows the effects of negative refraction to be observed. The following cases are considered: (a) scattering of plane waves by negative inclusions in the shape of a cylinder, (b) a plane parallel plate introduced in a positive background medium, (c) inclusions of a similar shape but composed of a positive material.

The problem about the scattering of a plane wave by a cylinder has an analytical solution. This problem was solved numerically using Eqs. (14.31) and (14.32). Subsequently, the field was calculated at a given distance from the cylinder centre

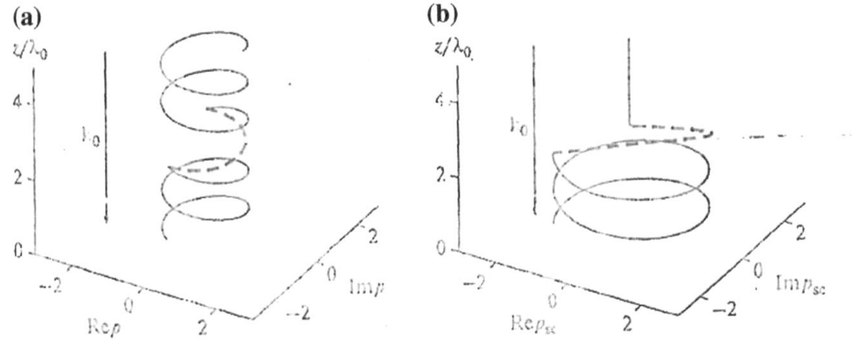

Fig. 14.7 Calculated total acoustic pressure field p (**a**) and scattered field p_{SC} (**b**) in a layer of a negative material. The solid and dashed lines indicate the fields in the background positive medium and in the layer placed in it, respectively. The arrow indicates the direction of the wave vector $\vec{k}_0$ of the incident wave (From Burov et al. [33])

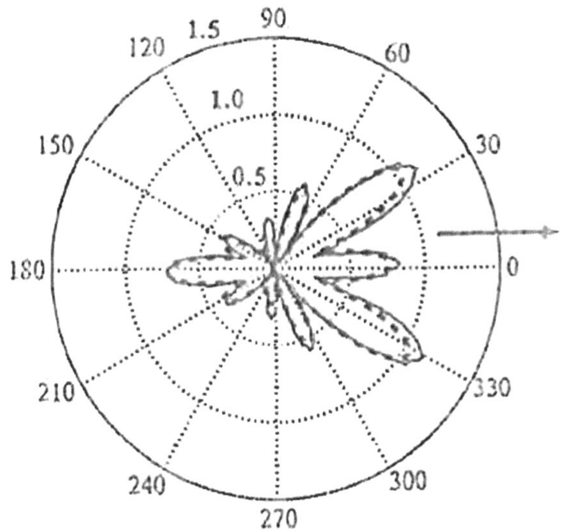

Fig. 14.8 Diagram of the acoustic pressure $|p_{SC}|$ of the field of a plane monochromatic wave with the unit's amplitude scattered by a cylinder of a positive material. The arrow indicates the direction of its propagation. The solid and dashed lines indicate the fields calculated using the exact formulae and Eqs. (14.31) and (14.32), respectively (From Burov et al. [33])

at various angles with respect to the direction of the incident wave. The results of the calculations for an inclusion in the shape of a cylinder of radius $R = \lambda_0$ with parameters $\rho \equiv 1$ and $\kappa \equiv 5$ of the positive material are presented in Fig. 14.8.

An angle of 0° corresponds to the direction of the incident plane wave in which the pressure amplitude was taken to be 1. The solid curve indicates the dependence of the pressure amplitude of the scattered field on the scattering angle at a distance of $4\lambda_0$ from the cylinder centre derived from the analytical solution, and the dotted line indicates the dependence calculated from Eqs. (14.31) and (14.32). The sampling step

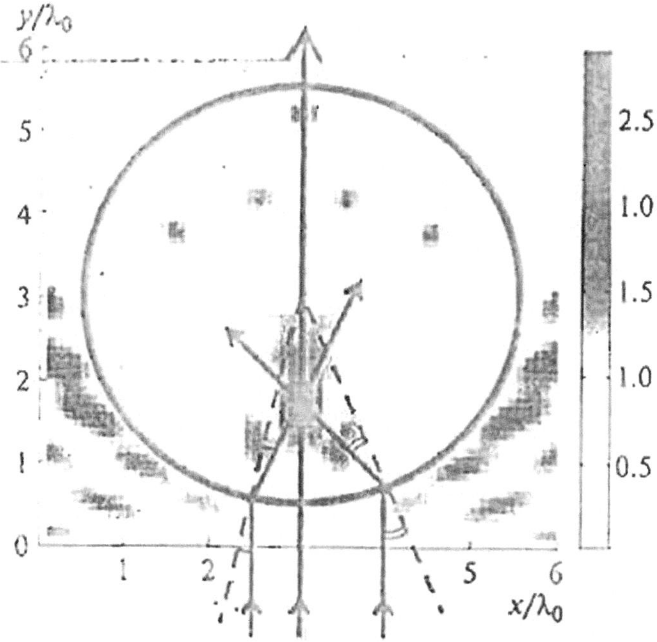

Fig. 14.9 Magnitude of the calculated acoustic pressure field $|p|$ of a plane wave refracted by a cylinder of a negative material. The arrows indicate the directions of the rays (From Burov et al. [33])

was $\lambda_0/10$. The deviation of the cylinder shape from a strictly cylindrical one caused by sampling in the calculation based on Eqs. (14.31) and (14.32) is responsible for the slight difference between the two curves. To within this error, a close coincidence is retained for various cylinder radii and contrasts. Thus, our comparison of the numerical simulations with the analytical solution indicates that the approach used is applicable.

In Fig. 14.9, the arrows indicate the path of the rays (corresponding to the pattern of energy propagation, i.e. the vector $\vec{S}$) calculated using Snell's law and their focusing in a cylinder of radius composed of a negative medium. For example, at $\rho \equiv -1$ and $\kappa \equiv -1$, a plane wave in the paraxial approximation is focused by the cylinder at a distance of r/z from its centre. The field for a cylinder with $r = 2.5\lambda_0$ is calculated numerically with a sampling step of $\lambda_0/10$. Although applying the geometric approximation for such small objects is not quite legitimate, the calculated total field in the cylinder also has a focus at this point, as can be clearly seen from Fig. 14.9.

To model the refraction of a wave on a plane parallel plate, a plane beam with a width of $5\lambda_0$ is used whose amplitude was taken to be 1. The plate has a thickness of $1.4\lambda_0$ and a length of $5\lambda_0$. It should be noted that since the method used does not allow scattering objects of infinitely large sizes or fragments of such objects to be modelled, this leads to an explicit or implicit appearance of boundary conditions

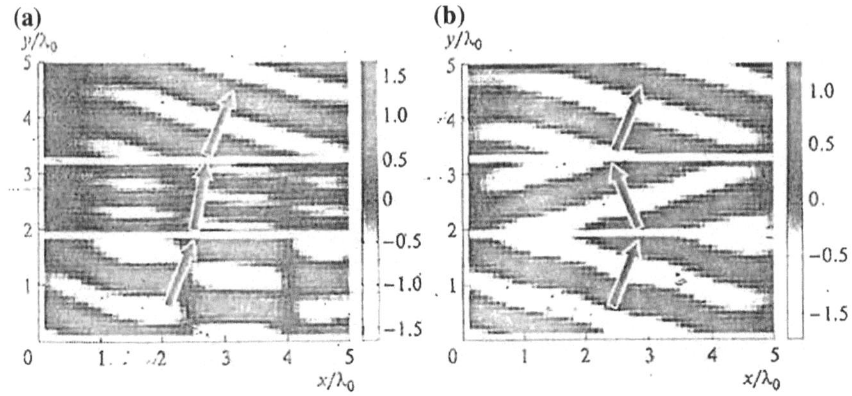

Fig. 14.10 Real part of the calculated acoustic pressure field $|p|$ for a plane wave incident on a plate of **a** positive and **b** negative materials. The arrows indicate the characteristic propagation directions of the wave energy; the white horizontal line represents the plate boundaries (From Burov et al. [33])

and unavoidable undesirable consequences in the form of false reflection and the formation of standing waves. To reduce the edge effects caused by the finite plate and beam sizes, the beam edges were artificially smoothed. The beam fell at an angle of 18° to the normal to plate.

Figure 14.10a shows the result of the calculation for a plate of a positive material ($\rho \equiv 1, \kappa \equiv 4$). The wavelength in it and, hence, the speed of sound are approximately half those in the background medium, in agreement with the formula for the speed of sound $c = 1/\sqrt{\rho\kappa}$. The normals to fronts of the incident and refracted waves lie on opposite sides relative to the normal to plate, which corresponds to the case of classical wave refraction at a boundary. The angles of incidence and refraction obey Snell's law. Since the impedances of the plate and the background medium do not match, there are waves reflected from the boundaries in the system that interfere and form a structure in the form of field minima and maxima along the front of the incident wave, as can be clearly seen from the figure.

When the beam falls on the plate of a negative material ($\rho \equiv -1$ and $\kappa \equiv -1$), the phenomenon of negative refraction is observed (Fig. 14.10b). The exact equality of the angles of incidence and refraction manifests itself in the symmetry of the picture of the wave fronts relative to the plate boundaries (the wavelength inside the plate remains equal to λ_0). The normal fronts of the incident and refracted waves directed along the vector $\vec{S}$ in the plate and the background medium are on the same side from the normal to plate. There is no reflected wave, since the impedances are equal. The vector $\vec{k}$ is directed along $\vec{S}$ in a positive medium and opposite to $\vec{S}$ in a negative one, and Snell's law is satisfied: the wave vectors of the incident and refracted waves have identical (not only in magnitude but also in sign) projections onto the interface. Thus, we have shown that a set of phenomena similar to the phenomena in left-handed media in electrodynamics follows from the simultaneous negativity of ρ and κ in an acoustic medium.

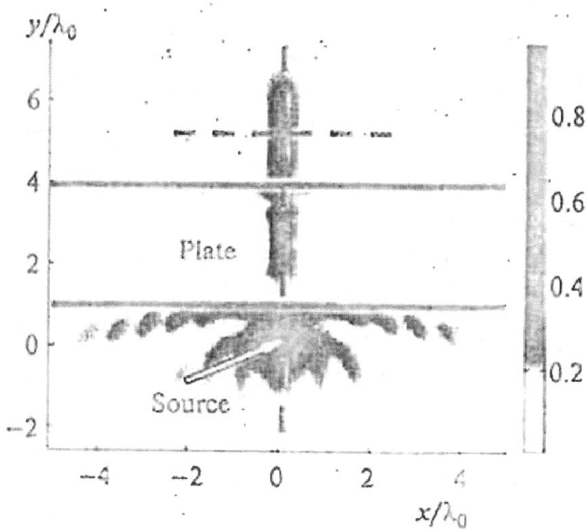

Fig. 14.11 Magnitude of the total calculated acoustic pressure field $|p|$, emerging when a point source is placed near a plane parallel plate of a negative material; the sampling step is $\lambda_0/10$. The dashes mark the lines along which the sections are considered; the solid horizontal lines represent the plate boundaries (From Burov et al. [33])

Next, the case of a negative medium perfectly matched with the background is considered. This is a case of special interest because a plane parallel plate of such a material is a focusing one (Fig. 14.11).

It was noted in [36] that such a lens has a number of peculiarities. First, as has already been illustrated, there are no losses by the reflection of the incident wave from it. Secondly, it has no focal plane. The image produced by it is a three-dimensional one, as in the case of a mirror, but, in contrast to the latter, it is real one. Thirdly, for each of rays emanated from one point, the length of the ray path is the same in negative and positive media. The coordinate phase shift in a negative medium is opposite in sign to that in a positive one, and as a result, they cancel each other out. Therefore, the wave phase at the image point is exactly equal to the wave phase at the corresponding source point. In this sense, a plane parallel plate of a negative material is a perfect lens. The equality of the phases at the source and image points does not lead to any paradox and does not violate the causality principle, since it takes place only at one frequency at which the effective density and compressibility of negative medium satisfy the relations $\rho = -\rho_0, \kappa = -\kappa_0$. Strict satisfaction of this condition is not possible even in a narrow frequency band, and this is probably an important restriction in the application of such lenses. Another shortcoming of such a lens is that it produces real images of objects located only at a short distance $L < H$ from it, where H is the plate thickness.

Such a lens was proposed theoretically by Pendry [30] in 2000 showing that Rayleigh diffraction limit can be defeated. Information about the source's fine details is contained in its near field. It includes the inhomogeneous waves that are usually neglected, since the decay exponentially with distance from the source, and are evanescent. However, the near field retains information about the initial phase and amplitude of the field near the source. Therefore, to reconstruct an ideal image, it will suffice to amplify it. A layer of a negative material can act as such an amplification. The fact that being a passive media the hypothetical ideal negative material has no energy source to amplify the evanescent waves appears paradoxical. However, this is not required: the decay of the evanescent field in a positive medium by increasing the distance from its source does not lead to any energy loss or heat release. By analogy, the inverse process in a negative medium does not require any expenditure of energy either.

The above shows that the results of the simulation for the refraction of the field of a point source illustrating the effect for plane parallel plates differing in length and parameters ρ, and κ. The thickness of all plates was the same, $3\lambda_0$. The scalar point source was located at the coordinate origin (0, 0) at a distance of $1.5\lambda_0$ from the plate and was specified in the form $F(\vec{r}) = \begin{bmatrix} \vec{f} = 0 \\ \rho = \delta(\vec{r}) \end{bmatrix}$. Its field was calculated at each point as $\check{u}_0 = \hat{\rho} * \check{F}$ and was normalized in such a way that the magnitude of the acoustic pressure at the source point was equal to 1.

A plate of a negative material ($\rho \equiv -1$ and $\kappa \equiv -1$) with a length of $10\lambda_0$ is considered. The distribution of the magnitude of the calculated pressure field $|p|$ in the chosen region is shown in Fig. 14.11. Two focal spots are clearly seen: in the middle of and behind the plate. When the negative refractive index of the plate is taken into account, their locations correspond to the geometric construction and confirm the picture of operation of a negative material.

To determine and investigate the resolution of the plate used as a lens, Burov et al. [33]. constructed two sections of the distribution of $|p|$: the longitudinal one (i.e. parallel to the normal to the plate) passing along the line connecting the centres of the source and the focal spots, and the transverse one (perpendicular to it) passing through the centre of the external focus (these sections are indicated in Fig. 14.11 by the dashes). The corresponding plots are presented in Fig. 14.12 (thick solid line). The values of $|p|$ in these sections were also calculated for a similar plate of negative material but with a length of $4\lambda_0$.

As the aperture of the lenses decreases, their resolution in both longitudinal and transverse directions is reduced (Fig. 14.12, thin solid line). For comparison, similar sections for a flat collecting lens with a length of $4\lambda_0$ are constructed. They are made in the shape of a plane parallel plate of a positive material with a variable (along the X-axis) refractive index [35]. The focus produced by it is much more blurred than that for a plate of negative material, particularly in the longitudinal direction (Fig. 14.12, dashed line). The increase in the resolution of the lenses of a negative material is related to the amplification of the evanescent field component of the source in the negative medium.

14.7.4 Points to Take Care of During Numerical Simulation

Choosing an optimal sampling step is important in a numerical analysis. Increasing the number of samples N per wavelength entails an increase in the sizes of the

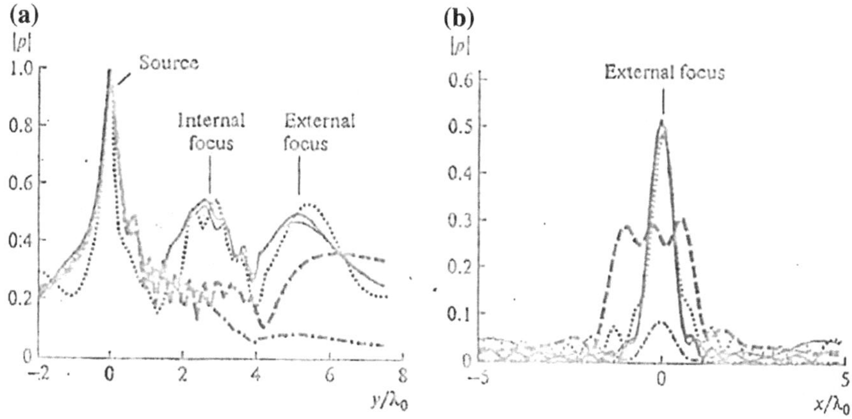

Fig. 14.12 Longitudinal (**a**) and transverse (**b**) sections of the acoustic pressure field passing through the external focus as shown in Fig. 14.11. The thick solid line corresponds to a lens of a negative material with a length of $10\lambda_0$, and the dash-dotted line corresponds to the same lens with absorption. The thin solid line corresponds to a lens of a negative material with a length of $4\lambda_0$; the dashed line corresponds to a lens of a positive material with a variable refractive index. In all these cases, the sampling step is $\lambda_0/10$. The dotted line corresponds to a lens of a negative material with a length of $10\lambda_0$ at a sampling step of $\lambda_0/7.5$ (From Burov et al. [33])

reversible matrices as $N^2 \times N^2$ (in the two-dimensional problem), increasing the amount of the random access memory used and the computational time. On the other hand, enlarging the resolution elements does not lead to any degradation of the quality of the field picture, possibly only to certain limits. In particular, as yet no increase in the sizes of the focal waist occurs (Fig. 14.12, dotted line). As the number of samples per wavelength decreases further, the field calculation procedure becomes unstable, causing the picture to be destroyed. The instability for plates of negative and positive materials arises at sampling steps of $\lambda_0/5$ or sparser and $\lambda_0/3$ or sparser, respectively. This is probably because when the evanescent field component is amplified, the errors caused by sparse sampling increase simultaneously. This is a manifestation of the fact that the problem here is an ill-posed one. The image quality can be slightly improved by applying regularization during the inversion of the operator in Eq. (15.7.5) similar to the least square method:

$$\check{u} = \left[\widehat{M}^{\dagger}\widehat{M} + \chi\widehat{E}\right]^{-1}\widehat{M} + \check{u}_0 \tag{14.34}$$

where $\widehat{M} \equiv \widehat{E} - \widehat{G} * \widehat{A}_1$, $\chi > 0$ is a small regularization parameter. The results of the calculation for a coarse sampling step of $\lambda_0/5$ for the plate with regularization are shown in Fig. 14.13. The image quality is restored but the focal spot broadens significantly. More surprisingly, the internal focus on the plate virtually disappeared and it is so far difficult to explain this fact.

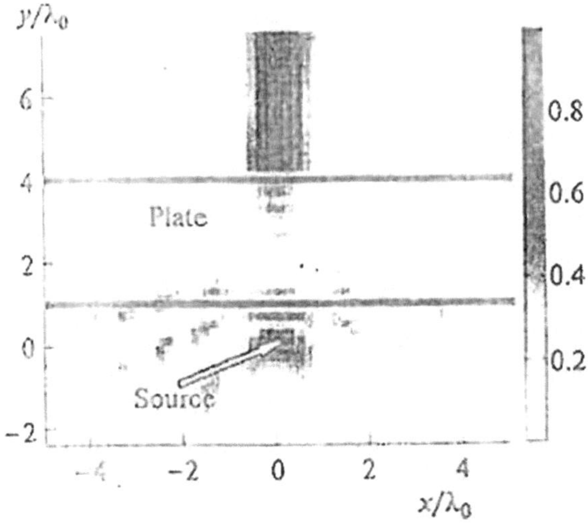

Fig. 14.13 Magnitude of the total acoustic pressure field $|p|$ emerging when a point source is placed near a plane parallel plate of a negative material and calculated using regularization. The sampling step is $\lambda_0/5$ (From Burov et al. [33])

It should be noted that choosing a sampling step is important not only in computer simulations but also for the practical creation of negative metamedia, which are discrete in their nature. In particular, to achieve an acceptable lens resolution, there should be at least ten discrete elements of the metamedium per wavelength. This requirement is actually reduced to the fact that the speed of sound inside the element of the resonating inclusion should be at least an order of magnitude lower than that in the background medium, which is significantly difficult in creating such media.

14.8 Extension to Theory of Diffraction by Inclusion of General Form of Mass Density and Bulk Modulus Manipulated by Predetermined Direction of Sound Propagation

In this section, we will extend Burov et al.'s [33] treatment of using double-negative media to the case of a general form for the mass density and the bulk modulus manipulated by predetermined direction of sound propagation. This will further illustrate that negative refraction for double-negative media is a special case of coordinate transformation. This shows that diffraction can be manipulated and controlled. This is the actual situation of sound propagation in a metamaterial. The theory will be developed along the line of Burov et al. [33], and this will be the further development of this chapter.

14.9 A New Approach to Diffraction Theory—A Rigorous Theory Based on the Material Parameters

So far, the usual theory of diffraction is based on the Kirchhoff diffraction formula which describes wave propagation encountering obstacles and formulates as a special form of scattering by apertures and boundary value problem with condition imposed by the relation between the wavelength and the size of the particle. It is applicable to both sound wave and electromagnetic wave because only general wave phenomenon is considered without differentiating between the mechanical properties of the sound wave and the electric and magnetic properties of electromagnetic wave.

Rigorous theory of diffraction for electromagnetic wave is given by Clemmow in a chapter on rigorous diffraction theory in the book Principles of Optics by Born and Wolf [39]. In this chapter, rigorous theory of diffraction of electromagnetic wave takes into account the polarization of the electric field and the magnetic field during diffraction. Examples of rigorous diffraction theory for electromagnetic wave are Mie's theory of scattering spheres given in 1908, Bouwkamp's dissertation [40], Wiener–Hopf method and H. Levine and J. Schwinger's application of variational principle to diffraction problem [41]. Here, we will use a new approach to rigorous diffraction theory for electromagnetic wave by considering the manipulation of the permeability and the permittivity giving rise to the control of the electromagnetic field. Since we are now able to bend or control the direction of light propagation to any predetermined direction, we in turn can control and manipulate the diffraction of light wave. This is one step further than the usual rigorous diffraction theory for electromagnetic wave given above because those theories only describe the behaviour of the light wave but here we are dealing with application of the outcome of the control and manipulation of the direction of propagation of the light wave.

We extend the idea to the sound wave. We are now able to manipulate the material parameters of the sound wave given by mass density and the bulk modulus using acoustical metamaterial. Hence, we are able to manipulate and control the propagation of the sound wave to any predetermined direction. This will give rise to the control and manipulation of diffraction. An example will be the diffraction of the sound wave by a screen made of acoustical metamaterial with an aperture.

14.10 Negative Refraction Derived from Reflection Invariance (Right–Left Symmetry)—A New Approach to Negative Refraction

This is another example of the gauge invariance approach to acoustic fields and is a more formal theory for negative refraction compared with the dispersion relation approach of Veselago [1]. Gauge invariance approach is a more generic approach as it covers both negative refraction and cloaking problem. The existence of symmetry laws is in full accordance with our daily experience. The invariance of physical laws under a coordinate transformation of uniform velocity, also known as the invariance under Galilean transformations, is a more sophisticated symmetry than was earlier recognized, and formed one of the cornerstones of Newtonian mechanics. Consequences of these show important results. A good example of the application of symmetry principles or gauge invariance to acoustic fields results in the theorem that during sound propagation in an isotropic solid gives rise to only two elastic constants.

Another consequence of the symmetry principles gives rise to the existence of conversation laws. For instance, the invariance of physical laws under space displacement gives rise to the conservation of momentum and the invariance under space rotation gives rise to the conservation of angular momentum. With advent of special and general relativity, the symmetry laws gained new importance.

Here, we interpret the negative refraction as the mirror image or the reflection of the usual positive refraction. The laws of physics have always shown complete symmetry between the left and the right. Hence, the acoustic equation on motion also possesses reflection invariance or right–left symmetry. The right–left symmetry is a discrete symmetry unlike rotational symmetry which is a continuous symmetry. In our previous Sect. 14.5, we have introduced the concept of parity. The left-handed system will correspond to parity equal to -1, and the right-handed system will correspond to parity equal to $+1$.

This can be further illustrated by the following rotation of the coordinate system about the vertical z-axis in the clockwise direction. Here, we will show that rotation of the coordinate system about the z-axis in the clockwise or left-handed direction for 180° will produce reflection invariance or mirror image by using the transformation matrix or the direction cosines matrix for the case of a left-handed rotation for 180° or inversion of the coordinate axis.

For the case of a left-handed rotation for 180° or inversion of the coordinate axis, the direction cosines matrix or coordinate transformation matrix can be written as follows:

$$\begin{pmatrix} a_{11} & a_{12} & a_{13} \\ a_{21} & a_{22} & a_{23} \\ a_{31} & a_{32} & a_{33} \end{pmatrix} = \begin{pmatrix} \cos 180^\circ & \cos 90^\circ & \cos 90^\circ \\ \cos 270^\circ & \cos 180^\circ & \cos 90^\circ \\ \cos 90^\circ & \cos 90^\circ & \cos 180^\circ \end{pmatrix} = \begin{pmatrix} -1 & 0 & 0 \\ 0 & -1 & 0 \\ 0 & 0 & -1 \end{pmatrix} \tag{14.35}$$

$$\begin{pmatrix} \varepsilon_{11} & \varepsilon_{12} & \varepsilon_{13} \\ \varepsilon_{21} & \varepsilon_{22} & \varepsilon_{23} \\ \varepsilon_{31} & \varepsilon_{32} & \varepsilon_{33} \end{pmatrix} \times \begin{pmatrix} -1 & 0 & 0 \\ 0 & -1 & 0 \\ 0 & 0 & -1 \end{pmatrix} = \begin{pmatrix} -\varepsilon_{11} & -\varepsilon_{12} & -\varepsilon_{13} \\ -\varepsilon_{21} & -\varepsilon_{22} & -\varepsilon_{23} \\ -\varepsilon_{31} & -\varepsilon_{32} & -\varepsilon_{33} \end{pmatrix} \tag{14.36}$$

$$\begin{pmatrix} \mu_{11} & \mu_{12} & \mu_{13} \\ \mu_{21} & \mu_{22} & \mu_{23} \\ \mu_{31} & \mu_{32} & \mu_{33} \end{pmatrix} \times \begin{pmatrix} -1 & 0 & 0 \\ 0 & -1 & 0 \\ 0 & 0 & -1 \end{pmatrix} = \begin{pmatrix} -\mu_{11} & -\mu_{12} & -\mu_{13} \\ -\mu_{21} & -\mu_{22} & -\mu_{23} \\ -\mu_{31} & -\mu_{32} & -\mu_{33} \end{pmatrix} \tag{14.37}$$

$$\begin{pmatrix} \rho_{11} & \rho_{12} & \rho_{13} \\ \rho_{21} & \rho_{22} & \rho_{23} \\ \rho_{31} & \rho_{32} & \rho_{33} \end{pmatrix} \times \begin{pmatrix} -1 & 0 & 0 \\ 0 & -1 & 0 \\ 0 & 0 & -1 \end{pmatrix} = \begin{pmatrix} -\rho_{11} & -\rho_{12} & -\rho_{13} \\ -\rho_{21} & -\rho_{22} & -\rho_{23} \\ -\rho_{31} & -\rho_{32} & -\rho_{33} \end{pmatrix} \tag{14.38}$$

$$\begin{pmatrix} \kappa_{11} & \kappa_{12} & \kappa_{13} \\ \kappa_{21} & \kappa_{22} & \kappa_{23} \\ \kappa_{31} & \kappa_{32} & \kappa_{33} \end{pmatrix} \times \begin{pmatrix} -1 & 0 & 0 \\ 0 & -1 & 0 \\ 0 & 0 & -1 \end{pmatrix} = \begin{pmatrix} -\kappa_{11} & -\kappa_{12} & -\kappa_{13} \\ -\kappa_{21} & -\kappa_{22} & -\kappa_{23} \\ -\kappa_{31} & -\kappa_{32} & -\kappa_{33} \end{pmatrix} \quad (14.39)$$

Equations (14.36) to (14.39) demonstrate that the values of the permittivity, permeability, mass density and bulk modulus will become negative after the left-handed rotation for 180° or inversion of the coordinate axis.

In fact, the $-\mu$ and $-\varepsilon$ can be considered as the mirror image of μ and ε and the $-\rho$ and $-\kappa$ can be considered as the mirror image of ρ and κ.

The direction cosines concept was used by Sokolov [26] and was in turn used and quoted by Veselago [1] as follows:

"A wave propagation in a given medium will be characterized by the matrix [26]:

$$\begin{pmatrix} \alpha_1 & \alpha_2 & \alpha_3 \\ \beta_1 & \beta_2 & \beta_3 \\ \gamma_1 & \gamma_2 & \gamma_3 \end{pmatrix}$$

The determinant of this matrix is equal to -1 if this set is left-handed [26]."

Further than this, Veselago [1] has to quote from Mandel'stam [28] for his concept of negative refraction:

"As we see, when the second medium is left-handed the refracted ray lies on the opposite side of the z axis from its position in the case of a right-handed second medium [28]."

Section 14.5 uses the gauge invariance approach as an alternative theory to negative refraction, and Sect. 14.10 further enhances this idea by applying the reflection invariance (or right–left symmetry) of acoustic wave equation to negative refraction. In so doing, the ambiguity of the choice of the negative sign instead of the positive sign in the Veselago's [1] dispersion relation of n = refractive index $= \pm\sqrt{\mu\varepsilon}$ is removed. Besides this, using coordinate transformation matrix we also demonstrated mathematically that for the case of negative refraction, it will also give rise to negative value of the permeability and negative value of the permittivity for the electromagnetic wave and negative mass density and negative bulk modulus for the acoustic wave. This presentation is clearer than that used by the Veselago paper [1] who used a qualitative argument.

14.11 A Unified Theory for the Symmetry of Acoustic Fields in Isotropic Solids, Negative Refraction and Cloaking

In section five, we have shown that both negative refraction and cloaking can be unified under the gauge invariance or coordinate transformation approach to acoustic fields. In isotropic solids, the acoustic fields are the same or invariant in all directions.

From the definition of the isotropic medium, it follows that its elastic properties are independent of the orientation of coordinate axes. In particular, elastic constant c_{ij} must remain invariant when we introduce new coordinate axes x_1', x_2', x_3' obtained by rotating the x_1, x_2, x_3-system through an angle about the x_1-axis. This is a form of rotational invariance or rotational symmetry. Due to rotational symmetry, the elastic constants in isotropic solids which is a fourth rank tensor, with 36 tensor components, can be tremendously simplified to only 2 components: c_{11} and c_{44}. This is an application of invariance in isotropic solids. The isotropy invariance, reflection invariance and time reversal invariance are all subsets of gauge invariance or invariance in coordinate transformation. Hence, we can unify the rotational symmetry of acoustic fields in isotropic solids, negative refraction and cloaking under the umbrella of coordinate transformation or gauge invariance. This amounts to the rediscovery of the symmetry nature of acoustic fields in solids.

14.12 Application of New Acoustics to Acoustic Waveguide

The ability to use double-negative (DNG) material to control and manipulate scattering will give rise to a new form of scattering and a new theory of scattering. The use of double-negative (DNG) material as substrate of acoustic waveguide can improve the directivity and optimize the far field radiation. Wu et al. [42]. studied the use of metamaterial as antenna substrate to enhance gain for electromagnetic waves. Their idea can be extended to acoustic waves. They studied the radiation set-up, simulated the farfield radiation, calculated the electric fields and radiated power. They manipulated ε and μ through specific inclusion of metal in dielectrics to achieve substrate properties in order to yield optimum radiation characteristics. For calculation of scattering in acoustical waveguides, the following procedure has to be taken [43].

To calculate the scattering of quasi-longitudinal plane sound wave at a boundary between two anisotropic media of different parities, one with parity equals to $+1$, and a normal double-positive (DPS) material and a left-handed (DNG) material with parity equal to -1. Auld [41] has obtained results for two media of the same parity $+1$. It would be of great interest to extend this calculation to two media of different parities. These will produce equations for new reflection coefficient, refraction coefficient, transmission coefficient and scattering coefficient for both SV wave and p wave for different polarizations. Fresnel equations for anisotropic solids for the two media of different parities will have to be derived. This will study the scattering at the boundary between the two media. It would be of great interest to study the behaviour of reflected evanescent wave and transmitted evanescent wave at the boundary of two media of different parities.

The new phenomena of scattering in DNG material can be manifested in the salient and conventional features of acoustic cavity resonators, acoustic waveguides, scatterers and antennas loaded or covered with double-negative and/or double-positive (DNG, DPS) metamaterials. The unconventional acoustical characteristics of metamaterials are exhibited when these materials are paired with other materials with at least one oppositely signed constitutive parameter. In other words, when we pair a DNG material with a DPS layer, we may obtain interesting wave propagation properties that may be absent if we paired one DNG layer with another one.

It is of great interest to notice that the interface between two media with at least one pair of oppositely signed parameters can play a major role in offering anomalous behaviours for the combined structure. At the boundary between such two media, using the acoustic field equation one can write the continuity of the tangential stress field and particle velocity field components. It is clear that the normal derivatives of these tangential components are not necessarily continuous and furthermore if ρ_1 and ρ_2 and/or κ_1 and κ_2 have opposite signs, then the derivatives of the tangential fields on both sides of the interface will have opposite signs. The discontinuity for the tangential components of fields at the interface between such media may imply a concentrated resonant phenomenon at that interface similar to the current and voltage distributions at the junction between an inductor and a capacitor at the resonance of an L-C circuit. This feature can lead to interesting characteristics for wave interaction in devices and components containing metamaterials.

It is also worth noticing that this interface resonance is independent of the total thickness of the paired layers, since it arises along the discontinuity between two such conjugate materials. The mechanism behind this resonance can be described by the equivalent circuit approach. This resonant characteristic which may occur in subwavelength structures formed by pairing such media has provided us with ideas for acoustic cavities, acoustic waveguides, scatterers, acoustic antennas and acoustic lenses that may operate with dimensions below the conventional diffraction limits.

This concentrated resonant phenomenon can be used to design, thin, subwavelength acoustic cavity resonators and parallel plate acoustic waveguides in which a layer of DNG material is paired with a layer of DPS material. By exploiting the antiparallel nature of the phase velocity and acoustic Poynting vector in a DNG slab, we found the possibility of resonant modes in electrically thin parallel plate structures containing such bilayered structures. This is an outcome of parity equal to -1.

14.13 New Elasticity

The ability to fabricate metamaterial producing negative permeability enables the beginning of new magnetism and the controlling and manipulation of electromagnetic fields. On the other hand, the ability to fabricate acoustical metamaterials with new physical properties of negative mass density and negative compressibility or bulk modulus marks the beginning of new elasticity enabling the controlling and the manipulation of the elastic fields or the elastodynamic fields. Our future works will be in this direction which will lead to several new phenomena in acoustics and the applications in novel acoustic devices.

14.14 Nonlinear Acoustics Based on the Metamaterial

14.14.1 Principles

This is an introductory of nonlinear acoustics based on metamaterial. The metamaterial chosen here is the nonlinear phononic crystals, a form of bandgap metamaterial. The wave dynamics in strongly nonlinear phononic crystals based on granular chains in a silicone elastometer or Teflon matrix will be considered. The wave equation for strongly nonlinear solitary wave has to be used [44]. This is more general than the weakly nonlinear KdV equation:

$$u_{tt} = -c^2\left\{(-u_x)^{3/2} + \frac{a^2}{10}\left[(-u_x)^{1/4}\left((-u_x)^{\frac{5}{4}}\right)_{xx}\right]\right\} \tag{14.40}$$

where $-u_x > 0$, $c^2 = \frac{2E}{\pi\rho_0(1-\upsilon^2)}$, $c_0 = \left[\frac{3}{2}\right]^{1/2} c\xi_0^{1/4}$.

Here, c is not a sound speed, and instead c_0 is a sound speed corresponding to initial strain ξ_0. This equation has no characteristic wavespeed independent on amplitude. Despite its complex nature, it has simple stationary solutions with unique properties. In a system moving with a speed υ_p, its periodic solution is represented by a sequence of humps ($\xi_0 = 0$) [45]:

$$\xi = \left[\frac{5\upsilon_p^2}{4c^2}\right]^2 \cos^4\left[\frac{\sqrt{10}}{5a}x\right] \tag{14.41}$$

Solitary shape can be taken as one hump of periodic solution (with only two harmonics) with finite length equal to five particle diameters. This unique wave was observed in numerical calculations and detected in experiments [46]. Solitary wave can be considered as a quasiparticle with mass equal about 1.4 mass of grain in the chain, and its speed V_s has a nonlinear dependence on maximum strain ξ_m or particle velocity υ_m:

$$\upsilon_s = \frac{4}{5}^{1/2} c\xi_m^{1/4} = \left[\frac{16}{25}\right]^{1/5} c^{4/5}\upsilon_m^{1/5} \tag{14.42}$$

(14.42) shows that the speed of this wave can be infinitely small if the amplitude is small. This means that using this material as a matrix in nonlinear tunable phononic crystals (NTPCs) one can ensure infinite elastic contrast of components, important for monitoring of bandgaps. At the same time, speed of solitary waves can be considered as constant at any relatively narrow interval of amplitudes due to power-law dependence with small exponent. These properties enable NTPC to be used as effective delay lines with exceptionally low speed of signal.

Also when one-dimensional testing was performed by Daraio et al. [44] using chains of stainless steel spheres placed in Teflon or silicone elastomer matrixes, a remarkable feature of "sonic vacuum" is observed, meaning a very rapid decomposition of initial impulse on the distances comparable with the soliton width. This example also demonstrates that short duration impact on highly nonlinear ordered periodic systems (lattices) with weak dissipation may result in a chain of solitary waves instead of intuitively expected shock wave. This property of strongly nonlinear phononic crystal can be used for controlled impulse transformation in relatively short transmission lines. If chains of grains are placed into a polymer matrix, the nonlinear elastic behaviour is accompanied by strong dependence of electrical resistivity on local pressure [46]. Nesterenko et al. [47] also performed testing of the heavy–light interface of the two strongly nonlinear granular media under the magnetically induced precompression resulted in a dramatic change of reflectivity. Anomalous reflected compression waves and transmitted rarefaction waves were detected in experiments and numerical calculations. They name this phenomenon the "acoustic diode" effect because of the dramatic change of the reflectivity triggered by the precompression. The nonlinear phenomenon described here can be used as tunable controllers of information flow through interfaces and in the design of novel types of tunable shock protection layers. The precompression can be employed for designing tunable information transportation lines with the unique possibility of manipulating the signal's delay, reflection and decompositions at will for security-related information.

It also has to be noted that the tunability of strongly nonlinear system with the variation of the elastic modulus of the stainless steel beads can be used for the design of tunable acoustic focusing lenses.

14.14.2 Nonlinear Acoustic Metamaterials for Sound Attenuation Applications

The previous section shows that an acoustic crystal composed of tightly packed spherical particles can exhibit a wide spectrum of acoustic properties with responses varying from linear to highly nonlinear regimes. The physical attractiveness of these crystals resides in the controllability of such acoustic responses by simple manipulation of static precompression applied to the material. Yang and Daraio [48] showed that from the fundamental understanding of the energy transmission through these crystals in relation to the tunable acoustic nonlinearity, one can control the degree of nonlinearity. To do this, three parameters have to be varied: precompression, striker velocity and striker mass, maintaining an identical configuration of one-dimensional granular structures. The transmission gain in the stop/pass frequency band of the granular chain was evaluated as a function of the nonlinearity. The evolution of the frequency band structure was studied as the degree of nonlinearity was changed. The transmission gain of the granular structure showed a remarkable dependence on the structural linearity level.

To combine the frequency filtering response governed by the discrete particles with an amplitude filtering response, they assembled a system composed of a highly nonlinear granular chain and a deformable linear medium. Acoustic wave propagation can then be efficiently manipulated and redirected with such an added degree of freedom.

In the second part of the study, they build a hybrid linear/nonlinear metamaterial to allow high-energy transmission only in a selected range of external impact amplitude. In this hybrid structure, the nonlinear granular chain takes the role of transmitting energy when the system is excited with low-amplitude impacts, whereas the linear structure performs as an effective shock mitigation medium under large-amplitude impacts, controlled by structural deflections. A strong correlation of transmission gain with external impact amplitudes was verified, showing an order-of-magnitude reduction of transmission gain for large-amplitude impacts compared to that of low-amplitude impacts. The wave propagation and impact mitigation were evaluated in the nonlinear acoustic metamaterial using a combined discrete particle model and a finite element method. Finally, it was verified that the numerical results are in excellent agreement with the experimental results.

The proposed metamaterials are fundamentally different from any other approach to vibrating isolation. They do not use active modulation to suppress vibration/impacts, but rely on passive insulation. Furthermore, they are stiff and load-bearing, present large recovery to external deformation and do not develop permanent damage in the ranges of excitations studied. The proposed systems are designed to forbid the propagation of waves in selected frequency ranges (also known as bandgaps or stopbands). Incident waves in these forbidden frequency ranges experience an exponential decay of their amplitudes (i.e. they are evanescent waves), and they are fully reflected. The presence of nonlinearity in the structure may allow the redirection of part of the incoming energy into allowed modes. This type of acoustic filter system could be useful in protecting soldiers against hearing damage from weapon fire, while largely retaining their situational awareness.

14.15 Ultrasonic Attenuation in Acoustic Metamaterial

In this section, we will be describing the mechanisms associated with local resonances that cause wave attenuation in acoustic metamaterials when the effective mass density becomes negative. Attention will be focused on the mechanism that prevents harmonic waves from propagating in the metamaterial when the wave frequency is near the local resonance frequency. This work was done by Huang and Sun [49]. To aid the understanding of this phenomenon, the work done on the lattice by an external force and the subsequent energy flow in the mass-in-mass lattice system is examined, especially at forcing frequencies that near the local resonance frequency. They also produced numerical results to illustrate the characteristic dynamic behaviour of an acoustic metamaterial.

14.15.1 Mechanism of Energy Transfer and Wave Attenuation

In the equivalent monatomic lattice of infinite extent, a negative effective mass indicates spatial attenuation of wave amplitude. Since the original lattice system is purely elastic, the energy carried by the wave cannot be dissipated. A similar observation has also been obtained by when they studied a nonlinear lattice model. Based on the principle of conservation of energy, the energy transferred through an elastic material must be conserved. Consequently, wave attenuation in a perfectly elastic medium implies that the energy must be transferred to and stored somewhere instead of propagating along the lattice system. The explanation can be found by considering the energy flow and transport velocity in the system. When no energy absorption is present, the velocity of energy transport is equal to the group velocity. In order to understand the mechanism that causes the spatial attenuation of wave amplitude within the bandgap and to facilitate the use of numerical methods, they consider a lattice model with a finite number of mass-in-mass units. At one end of the finite lattice model, a forcing agent is applied providing continuous energy input to the lattice in the form of propagating waves.

To understand the work done by an external force on a lattice system, they first consider a single undamped mass-in-mass oscillator. The oscillator is connected to a rigid wall by a linear spring. It is found that negative work is done (i.e. energy is taken out of the system) during half of the excitation cycle. The behaviour of the negative work done will be investigated next in the finite lattice model later to explain the bandgap effect on wave propagation. They next consider the mass-in-mass lattice system. The finite element method is used to perform a numerical study of wave propagation in a lattice with 2,500 lattice units. The lattice spacing is chosen to be sufficiently smaller than the longitudinal wavelength. The selected excitation is the prescribed harmonic displacement. It is noted that the forcing frequency which is still in the range of negative effective mass is not so close to the local resonance, and the wave attenuation would take a longer distance to complete. As a result, the total input work does not return to zero in the initial cycles indicating that some amount of energy is leaked into the system in the form of attenuated wave propagation and would take a longer time to stop the wave. They conclude firstly that while in the bandgap, the propagating wave is attenuated because the energy is taken out by the driving force in the form of negative work. Secondly, when close to the local resonance, a large amount of energy is temporarily stored in the internal mass and later still taken out by the driving force.

14.15.2 Applications

An application of the above new mechanism of ultrasonic attenuation in acoustic metamaterial will be in metamaterials with tailorable material constants. A possible metamaterial with negative effective mass density may be constructed in the form of composite material consisting of periodic mass-in-mass microstructures embedded in a matrix material. The metacomposite possesses a local resonance frequency, an essential element to the desired metamaterial. The most characteristic of metamaterials is the local resonance frequency which can be easily tailored by choosing a softer spring or by using a larger internal mass. This means that the range of bandgap frequencies can be shifted at ease, making this a potential material for blocking low-frequency sound and vibration.

They also consider spatial wave attenuation in acoustic metamaterial by choosing free harmonic wave propagation in a mass-in-mass lattice of infinite extent. When the wave frequency approaches the local resonance, the propagating wave exhibits significant attenuation as expected. It is clear that the mass-in-mass microstructure is capable of stopping a wave from propagating into the metamaterial if the wave frequency is near the local resonance frequency. For a pulse consisting of a spectrum of frequencies, a number of microstructures with different local resonance frequencies are needed. As an example, they consider a metamaterial with four sections of different microstructures (or mass-in-mass units). Each section consists of 500 unit cells with a specifically tuned local resonance. The four resonance frequencies are artificially selected close to but lower than 100, 200, 300 and 400, respectively. Thus, the total metamaterial is formed by 2,000 mass-in-mass units of four different local resonance frequencies.

The excitation is generated by a prescribed displacement given by

$$u(0,t) = 0.01(\sin 100t + \sin 200t + \sin 300t + \sin 400t)H(t) \tag{14.43}$$

where $H(t)$ = unit step function applied at one end of the lattice system with the other end fixed. It is easy to see that the generated wave is composed of four excitation displacement histories of four distinct frequencies. It is expected that the wave generated by $\sin 100^t H(t)$ would be attenuated within the first section layer and that generated by $\sin 200^t H(t)$ filtered out within the second section, and so on. They have shown that the combined wave almost disappears when propagating into the fourth section of the model. By analysing the frequency spectrum of the wave in each section, one can see that the input disturbances are filtered out section by section and nearly disappear in the last section.

Huang and Sun [49] studied the wave attenuation mechanisms in a mass-in-mass metamaterial containing locally resonant microstructures. They showed that this metamaterial with negative effective mass density demonstrates a new bandgap created by the presence of the local resonance. With the local resonance-type bandgap, the metamaterial behaves dynamically like a phononic bandgap metamaterial. The main advantage of the present metamaterial is that it is easy to design the stop-band location by tailoring the local resonance frequency. It was also shown that present metamaterial can give rise to a significant wave attenuation effect near the local resonance frequency. This property can be used to block waves from passing this metamaterial. They have demonstrated that a metamaterial containing multiple microstructures with a spectrum of local resonance frequencies provides the capability of drastically reducing the magnitude of stress waves generated by dynamic sources.

14.16 Applications of Phononic Crystal Devices

Recently, phononic crystals and devices have been scaled to VHF (30–300 MHz) frequencies and beyond by utilizing microfabrication and micromachining technologies. Microphononic crystal devices realized in low-loss solid materials have potential applications in radio frequency communications and acoustic imaging for medical ultrasound and nondestructive testing. Focusing devices made with phononic crystals could miniaturize acoustic lenses, adapt impedance and decouple the transducer size from the aperture [50]. Phononic crystal slabs and devices such as waveguides [51] and cavities [52] have been demonstrated in the very high frequency (VHF) band (30–300 MHz).

While VHF phononic crystals have communication applications in oscillators and intermediate frequency (IF) filters, in order to be widely applied to RF signal processing, phononic crystal devices must be scaled to relevant consumer wireless communication frequencies spanning from 850 MHz to 2 GHz, which includes the GSM, CDMA and GPS bands. Olsson III et al. [53] have reported phononic crystal slabs centred at 943 MHz with bandgap widths of 416 MHz, wide enough to cover both the GSM-850 and the GSM-900 cellular phone bands. The phononic bandgap is well predicted by the finite difference time domain (FDTD) model except for the region between 600 and 700 MHz where the measured bandgap closes but the FDTD bandgap remains open. The high-frequency closing of the bandgap is accurately modelled by FDTD as is much of the acoustic structure in this region.

In order for phononic crystal devices to be widely applied in RF communication applications, the losses introduced by the devices must be negligible compared to that of the couplers; i.e. the normalized loss must approach 0 dB. Phononic crystals centred at 943 MHz are formed by arranging 1.4 μm diameter W rods in a square lattice with a pitch of 2.5 μm inside a 1.85 μm thick suspended SiO_2 membrane. The resulting phononic crystal has a bandgap width of 416 MHz or 44% and a maximum bandgap depth of 35 dB. Waveguide devices formed by placing defects in the phononic lattice have also been realized with propagation frequencies of 780 and 1,060 MHz.

14.17 Comparison of the Significance of Role Played by Gauge Theory and Multiple Scattering Theory in Metamaterial—A Sum-up of the Theories of Metamaterial

Metamaterial can be divided into two main categories: the double-negative (DNG) and the bandgap (BDG) metamaterial. The theory covering the DNG material is the gauge invariance theory. Veselago [1] proposed using the dispersion relation to explain negative permeability and negative permittivity. This is not a full-fledged theory like the gauge theory. The gauge theory is a more generic theory than the theory based on the dispersion relation. The gauge invariance theory can be applied to perfect lens, sound reduction and acoustical cloaking. The theory for the bandgap metamaterial or phononic crystals is the multiple scattering theory (MST) [18]. The phononic crystals can be applied to perfect lens and sound and vibration reduction but not to acoustical cloaking. Hence, the gauge theory has a wider scope of applications than the multiple scattering theory.

Also, the gauge invariance approach to acoustic field has resulted in widespread applications. For instance, the application of invariance or symmetry principles to sound propagation in isotropic solids has resulted in the famous theorem that in an isotropic solid there are only two elastic constants, c_{11} and c_{44}. The application of reflection invariance or right–left symmetry has resulted in the famous negative refraction phenomenon, and the application of time reversal invariance of the acoustic equation of motion has resulted in the widespread applications of time reversal acoustics. This shows that gauge invariance approach is of all time more powerful than the multiple scattering theory (MST).

The strength of gauge theory can be further strengthened by what is shown earlier in this chapter that negative refraction is a special case of cloaking when the wave propagation path is linear and the determinant of the direction cosines matrix is -1 compared with the nonlinear bending wave propagation path of cloaking. The gauge invariance of the form of wave equation subjected to coordinate transformation is the unified theory for negative refraction and cloaking. This is a pattern of nature and is applicable to all equations of physics including Maxwell's equations and acoustic equation of motion.

14.18 Conclusions

New acoustics based on metamaterial is an emerging field. Basically, this is rewriting the whole field of acoustics. Further topics of new acoustics which are not covered in this chapter but will be fulfilled in future edition of this book are new acoustics in underwater acoustical imaging, seismic imaging on the application of seismic metamaterial to the protection of buildings from earthquakes and tsunamis and on the application of metamaterial in nonlinear acoustics such as to the cancellation of high intensity noise and to rewrite the topics covered by B. A. Auld's books on Acoustic Fields and Solids [43] such as on Christoffel equation for double-negative media, transmission line model for negative media, acoustic waveguides and acoustic resonators for negative media. In 2007 W.S.Gan [54] proposed gauge invariance approach to acoustic fields and rewrites the whole field of acoustics. This seems to realize in reality. This will involve the rederivation of the solutions of acoustic equations involving negative mass density and negative compressibility. Auld's [43] books are based on a right-handed system, and it will be of great interest to acousticians if they are rewritten based on a left-handed system.

References

1. Veselago, V.G.: The electrodynamics of substance with simultaneously negative values of permittivity and permeability. Sov. Phys. Usp. **10**(4), 509–514 (1968)
2. Pendry, J.B., Holden, A.J., Robbins, D.J., Stewart, W.J.: Magnetism from conductors, and enhanced non-linear phenomena. IEEE Trans. Microwave Theory Tech. **47**(11), 2075–2084 (2009)
3. Smith, D.R., Padilla, W.J., Vier, D.C., Nemat-Nasser, S.C., Schultz, S.: Composite medium with simultaneously negative permeability and negative permittivity. Phys. Rev. Letts. **84**, 10990641 (2000)
4. Li, J., Chan, C.T.: Double-negative acoustic material. Phys. Rev. E **70**, 055602 (R) (2004)
5. Chang, L.L., Esaki, L.: Semiconductor quantum heterostructures. Phys. Today **45**, 36 (1992)
6. Yablonovitch, E.: Inhibited spontaneous emission in solid- state physics. Phys. Rev. Lett. **58**, 2059 (1987)
7. Sheng, P. (ed.): Scattering and Localization of Classical Waves in Random Media. World Scientific, Singapore (1990)
8. Sigalas, M.M., Economou, E.N.: Elastic and acoustic wave band structure. J. Sound Vib. **158**, 377 (1992)
9. Kushwaha, M.S., Halevi, P., Dobrzynski, L., Djafari-Rouhani, B.: Acoustic band structure of periodic elastic composites. Phy. Rev. Lett. **71**, 2022 (1993)
10. Sigalas, M., Economous, E.N.: Band structure of elastic waves in two-dimensional array of rigid cylinders. Solid Sate Commun. **86**, 141 (1993)
11. Kushwaha, M.S., Haleri, P., Martinez, G., Dobrzynski, L., Djafari-Rouhani, B.: Theory of acoustic band structure of periodic elastic composites. Phys. Rev. B **149**, 2313–2322 (1993)
12. Kushwaha, M.S., Halevi, P.: Band-gap engineering in periodic elastic composites. Appl. Phys. Lett. **64**, 1085–1090 (1994)
13. Economou, E.N., Sigalas, M.M.: Elastic and acoustic wave band structure. J. Acoust. Soc. Am. **95**, 1735 (1994)
14. Sánchez-Pérez, J.V., Caballero, D., Mártinez-Sala, R., Rubio, C., Sánchez-Dehésa, J., Meseguer, F., Llinares, J., Gálvez, F.: Sound attenuation by a two-dimensional array of rigid cylinders. Phys. Rev. Lett. **80**, 5325 (1998)
15. Mártinez-Sala, R., Sancho, J., Sánchez, J.V., Gomez, V., Llinres, J., Meseguer, F.: Sound attenuation by sculpture. Nature (London) **378**, 241 (1995)
16. Montero de Espīnosa, F.R., Jimeńez, E., Torres, M.: Ultrasonic band gap in a periodic two-dimensional composite. Phys. Rev. Lett. **80**, 1208 (1998)
17. Torres, M., Montero de Espīnosa, F.R., Darcía-Pablos, D., García, N.: Sonic band gaps in finite elastic media: surface states and localization phenomena in linear and point defects. Phys. Rev. Lett. **82**, 3054 (1999)
18. Kafesaki, M., Economou, E.N.: Multiple scattering theory for three-dimensional periodic acoustic composites. Phys. Rev. B **60**, 11993–12001 (1999)

19. Korringa, J.: On the calculation of the energy of a Bloch wave in a metal. Physica (Amsterdam) **XIII**, 392 (1947)
20. Kohn, W., Rostoker, N.: Solution of the Shrödinger equation in periodic lattices with application to metallic lithium. Phys. Rev. **94**, 1111 (1951)
21. Ashcroft, N.W., Mermin, D.N.: Solid State Physics. Holt, Rinehart and Winston, New York (1976)
22. Economou, E.N.: Green's Functions in Quantum Physics. Springer, Berlin (1983)
23. Economou, E.N., Sigalas, M.M.: Stopband for elastic waves in periodic composite materials. J. Acoust. Soc. Am. **95**, 1734 (1994)
24. Kafesaki, M., Economou, E.N.: On the dynamics of locally resonant sonic composites. Phys. Rev. B **52**, 13317 (1995)
25. Zhang, X., Liu, Z.: Negative refraction of acoustic waves in two-dimensional phononic crystals. Phys. Lett. **85**, 341–343 (2004)
26. Sokolov, A.V.: Opticheskie Svoistva Metallov [The Optical Properties of Metals]. Fizmatgiz, Moscow (1961)
27. Borisenko, A.I., Tarapov, I.E.: Vector and Tensor Analysis. Dover, New York (1968)
28. Mandel'stam, L.I.: Group velocity in a crystal lattice. JETP **15**, 475 (1945)
29. Li, J., Lee, F., Yin, X., Bartal, G., Zhang, X.: Experimental demonstration of an acoustic magnifying hyperlens. Nat. Mater. **8**, 931–934 (2009)
30. Pendry, J.B.: Negative refraction makes a perfect lens. Phys. Rev. Lett. **85**, 3966 (2000)
31. Hawking, S., Mlodinow, L.: The Grand Design. Bantam Books, New York, pp 39 (2010)
32. Wang, W., Lin, L., Yang, X., Cui, J., Du, C., Luo, X.: Design of oblate cylindrical perfect lens using coordinate transformation. Optic Express **16**(11), 8094–8105 (2008)
33. Burov, V.A., Dmitriev, K.V., Sergeev, S.N.: Acoustic double-negative media. Acoust. Phys. **55**, 298–310 (2009)
34. Burov, V.A., Vecherin, S.N., Rumyantseva, O.D.: Statistical estimation of the spatial spectrum of secondary sources. Akust. Zh. **50**, 14 [Acoust. Phys. **50**, 9] (2004)
35. Voitovich, N.N., Katsenelenbaum, B.Z., Sizov, A.N.: Generalized Method of Eigenoscillation in Diffraction Theory. Nauka, Moscow (1977) (in Russian)
36. Barkhatov, A.N., Gorskaya, N.V., Goryunov, A.A., et al.: Acoustics in Problems. Nauka, Fizmatlit, Moscow (1996) (in Russian)
37. Bliokh, K.Y., Bliokh, Y.P.: Electrodynamic analysis of a Pendry lens. Usp. Fiz. Nauk **174**, 439 [Phys. Usp. **47**, 393] (2004)
38. Burov, V.A., Dimitriev, K.V., Sergeev, S.N.: Wave effects in acoustic media with a negative refractive index. Izv. Ross. Akad Nauk Ser. Fiz. **72**, 1695 (2008)
39. Born, M., Wolf, E.: On the theory of diffraction by an aperture in an infinite plane screen I and II. Principles of Optics, 4th edn. pp. 556–592. Pergamon Press, UK (1970)
40. Bauwkamp, C.J.: On the diffraction of an electromagnetic wave through a plane screen, Dissertation, Groningen (1941)
41. Levine, H., Schwinger, J.: Phys. Rev. **74**, 958 (1948); **75**, 1423 (1949)
42. Wu, B.I., Wang, W., Pacheco, J., Chen, X., Grzegorczyk, T., Kong, J.A.: A study of using metamaterials as antenna substrate to enhance gain. Prog. Electromagnet. Res. PIER **51**, 295–328 (2005)
43. Auld, B.A.: Acoustic Fields and Waves in Solids, vols. I and II, pp. 1–220. Robert E. Krieger Publishing Company, Malabar, Florida, USA (1990)
44. Daraio, C., Nesterenko, V., Jin, S.: Strongly nonlinear waves in 3D phononic crystals. In: Furnish, M.D., Gupta, Y.M., Forbes, J.W. (eds.) Shock Compression of Condensed Matter, pp. 197–200. American Institute of Physics, Maryland (2003)
45. Nesterenko, V.F.: Dynamics of Heterogeneous Materials. Springer, New York (2001)
46. Jin, S., Tiefel, T.H., Wolfe, R., Sherwood, R.C., Mottine, J.J.: Optically transparent, electrically conductive composite medium. Science **255**, 446 (1992)
47. Nesterenko, V.F., Daraio, C., Herbold, E.B., Jin, S.: Anomalous wave reflection from the interface of two strongly nonlinear granular media. Phys. Rev. Lett. **95**, 158702 (2005)
48. Yang, J., Daraio, C.: http://www.embeddedtechmag.com/component/content/article/12254

49. Huang, H.H., Sun, C.T.: Wave attenuation mechanism in an acoustic metamaterial with negative effective mass density. New J. Phys. **11**, 013003–013011 (2009)
50. Olsson III, R.H., El-Kady, I.: Microfabricated phononic crystal devices and applications. Meas. Sci. Technol. **20**(1), 012002 (2009)
51. Olsson III, R.H., El-Kady, I.F., Su, M.F., Tuck, M.R., Fleming, J.G.: Microfabricated VHF acoustic crystals and waveguides. Sens. Actuators, A **145**, 87–93 (2008)
52. Mohammadi, S., Eftekhar, A.A., Hunt, W.D., Adibi, A.: High-Q micro-mechanical resonators in two-dimensional phononic crystal slab. Appl. Phys. Lett. **94**, 051906 (2009)
53. Olsson III, R.H., Griego, S.X., El-Kady, I., Su, M., Soliman, Y., Goettler, D., Leseman, Z.: Proceedings of 2009 IEEE International Ultrasonics Symposium, pp. 1150–1153 (2009)
54. Gan, W.S.: Gauge invariance approach to acoustic fields. In: Akiyama, I. (ed.) Acoustical Imaging, vol. 29, pp. 389–394. Springer, Netherlands (2007)

The manufacturer's authorised representative in the EU is Springer Nature Customer Service Centre GmbH, Europaplatz 3, 69115 Heidelberg, Germany. If you have any concerns regarding our products, please contact ProductSafety@springernature.com

Printed and bound by CPI Group (UK) Ltd, Croydon, CR0 4YY
31/07/2026
02182873-0012